AF543348

CHEMISTRY OF GEOMETRICAL AND OPTICAL ISOMERISM

CHEMISTRY OF GEOMETRICAL AND OPTICAL ISOMERISM

Arjun Singh Negi

CYBER TECH PUBLICATIONS
4264/3, Ansari Road, Daryaganj, New Delhi-110002 (India)
Ph.: 011-23244078, 011-43559448 Fax: 011-23280028
E-Mail: cyberpublicationsdelhi@yahoo.com
Website: www.cybertechpublications.com

CHEMISTRY OF GEOMETRICAL AND OPTICAL ISOMERISM

Arjun Singh Negi

© Reserved

First Edition 2010

All rights reserved. No part of this publication may be reproduced, stored in a retrieval system or transmitted, in any form or by any means, electronic, mechanical, photo copying, recording or otherwise, without the prior permission of the publisher.

Published by : Asian Offset Printers

G.S. Rawat for **Cyber Tech Publications**
4264/3, Ansari Road, Daryaganj, New Delhi-110002 (India)
Ph.: 011-23244078, 011-43559448 Fax: 011-23280028
E-Mail: cyberpublicationsdelhi@yahoo.com
Website: www.cybertechpublications.com

PREFACE

This books has been specially written to the requirements syllabus of the students of all Indian Universities. The Subject matter has been discussed in such a simple way that the students will find no difficulty to understand it. Each chapter of this book contains complete theory and fairly large number of solved examples. The proofs of various principles and examples has been given with minute details. The subject matter has been presented in a simple and systematic manner.

To familiarize the student about expected type of questions in the examinations, important summary and review questions have been provided at the end of each chapter.

The author will feel amply rewarded if the book serve the purpose for which it is meant.

— *Author*

Contents

1

Configuration of Geometrical Isomers

INTRODUCTON

There is no general method for determining the configuration of geometrical isomers. In practice one uses a number of different methods, the method used depending on the nature of the compound in question. At the same time, the use of several methods, if applicable, will give more reliable results.

The following are methods which may be used mainly for compounds that owe their geometrical isomerism to the presence of a double bond, but several of the methods are special to geometrical isomers possessing a cyclic structure.

METHOD OF CYCLISATION

Wislicenus was the first to suggest the principle that intramolecular reactions are more likely to occur the closer together the reacting groups are in the molecule. This principle appears always to be true for reactions in which rings are formed, but does not hold for elimination reactions in which a double (or triple) bond is produced.

(a) Of the two acids maleic and fumaric, only the former readily forms a cyclic anhydride when heated; the latter does not form an anhydride of its own, but when strongly heated, gives maleic anhydride. Thus (I) is maleic acid, and (II) is fumaric acid.

Cyclisation reactions must be performed carefully, since one isomer may be converted into the other during the cyclising process, and so lead to unreliable results. In the above reaction, somewhat vigorous conditions have been used; hence there is the possibility that interconversion of the stereoisomers has occurred. Since maleic acid cyclises readily, and fumaric acid only after prolonged heating, the former is most probably the cis-isomer, and the latter the trans which forms maleic anhydride via the formation of maleic acid.

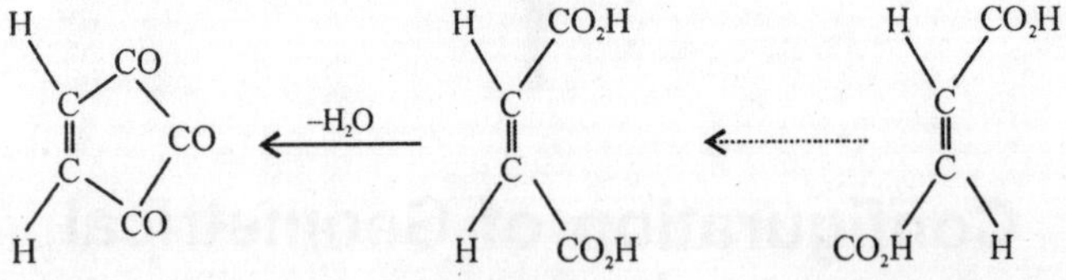

(I) maleic acid   (II) fumaric acid

The correctness of the conclusion for the configurations of the two acids may be tested by hydrolysing maleic anhydride in the cold; only maleic acid is obtained. Under these mild conditions it is most unlikely that interconversion occurs, and so we may accept (I) as the configuration of maleic acid.

(b) Citraconic acid forms a cyclic anhydride readily, whereas the geometrical isomer, mesaconic acid, gives the same anhydride but much less readily. Thus these two acids are:

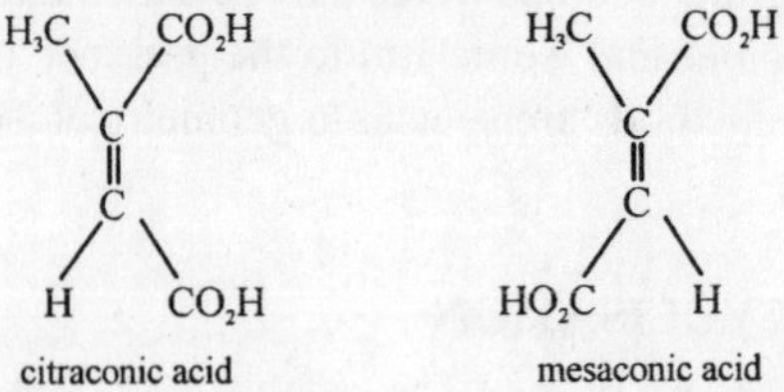

citraconic acid   mesaconic acid

METHOD OF CONVERSION INTO COMPOUNDS OF KNOWN CONFIGURATION

In a number of cases it is possible to determine the configurations of pairs of geometrical isomers by converting them into compounds, the configurations of which are already known.

As an example of this type let us consider the two forms of crotonic acid, one of which is known as crotonic acid (m.p. 72°C), and the other as isocrotonic acid (m.p. 15.5°C). Now there are two trichlorocrotonic acids, (III) and (IV), one of which can be hydrolysed to fumaric acid. Therefore this trichlorocrotonic acid must be the trans-isomer, (III);

consequently the other is the cis-isomer (IV). Both these trichlorocrotonic acids may be reduced by sodium amalgam and water, or by zinc and acetic acid, to the crotonic acids, (III) giving crotonic acid (V) and (IV) giving isocrotonic acid (VI). Thus crotonic acid is the-trans-isomer, and isocrotonic the cis (von Auwers *et al.,* 1923).

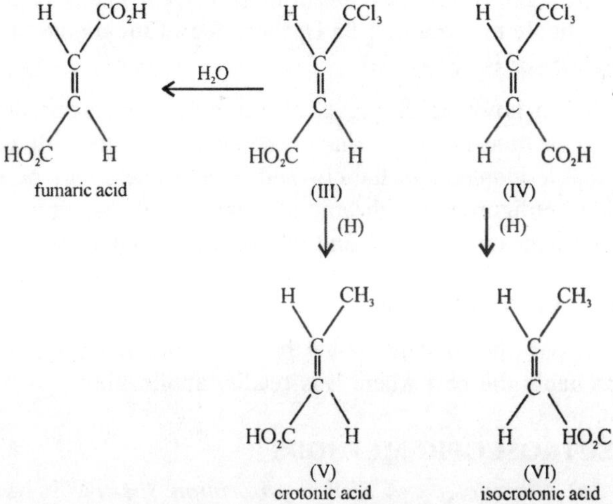

Another example is the reduction of crotonaldehyde (known to be the trans- (seqtrans or E-) form) into trans- (segtrans or E-) crotyl alcohol.

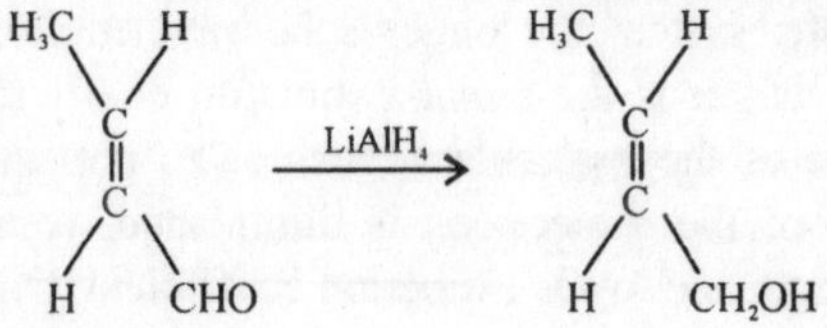

Method of Conversion into Less Symmetrical Compounds

Certain pairs of geometrical isomers may be converted into less symmetrical compounds in which the number of geometrical isomers is increased, and by considering the number of products obtained from each original stereoisomer, it is possible to deduce the configurations of the latter, *e.g,* there are two 2,5-dimethylcyclopentane1, 1-dicarboxylic acids, and these, on heating, are decarboxylated to 2,5-dimethylcyclopen tane1-carboxylic acid. The cis-form of the original dicarboxylic acid can give rise to two stereoisomeric monocarboxylic acids, whereas the trans-form can produce only one product. Thus. the configurations of the dicarboxylic acids are determined.

Method of Dipole Moments

The use of dipole moments to assign configurations to geometrical isomers must be used with caution.

The method is satisfactory so long as the groups attached to the olefinic carbon atoms have linear moments, *e.g,* cis-l,2-dichloroethylene has a dipole moment of 1.85 D; the value of the dipole moment of the trans isomer is zero.

When, however, the groups have non-linear moments, then the vector sum in the trans-isomer will no longer be zero and the difference between the dipole moments of the cis- and trans-isomers may be too small to assign configuration with any confidence, *e.g,* the dipole moment of diethyl maleate is 2.54 D and that of diethyl fumarate is 2.38 D.

X-ray Analysis Method

This method of determining the configuration of geometrical isomers is probably the best where it is readily applicable.

SPECTROSCOPIC METHODS

(a) *Ultraviolet and Risible Absorption Spectra:* It has previously been pointed out that absorption in compounds containing conjugation is due to $\pi \rightarrow \pi^*$ transitions, and that the longer the conjugated system, the longer is the wavelength of the absorption and the larger is the molar extinction coefficient. If, then, the structure of the molecule is such as to prevent planarity, the-overlap of the π-electrons is diminished, resulting in shorter wavelength and lower extinction coefficient. One factor that can decrease overlap is the steric factor, and since this would be larger in the cis-isomer than in the trans-, the latter would be expected to have the higher λ_{max} and ε.

A number of resonating structures are possible for both forms, and in all cases the C–C bond will have partial double-bond character, and consequently the molecule will tend to be planar. However, in the case of the cis-form, owing to the proximity of the two (large) benzene rings, there will be steric hindrance, resulting in decreased resonance, *i.e.,* decreased overlap of the it-electrons due to steric inhibition of resonance. This argument is supported by the fact that λ_{max} and ε for cis-stilbene are 278 nm (9 350), and for trans-stilbene are 294 nm (24 000).

(b) *Infrared Absorption Spectra:* Absorption brought about by = C–H bending is much more intense than that brought about by C=C stretching, and cis- and trans-isomers of the type shown may be distinguished by the different ν_{max} observed for the =C–H bending.

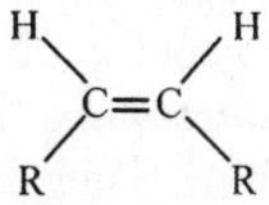
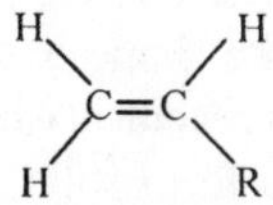

ν_{max} 730-665 cm^{-1} 970-960 cm^{-1}

(c) *NMR spectra:* The use of NMR spectroscopy for distinguishing between cis- and transisomers of the type CHa = CHa is based on the fact that the two hydrogen atoms have different coupling constants in each compound (see Table 1.9). This method may also be used to distinguish geometrical isomers of the type:

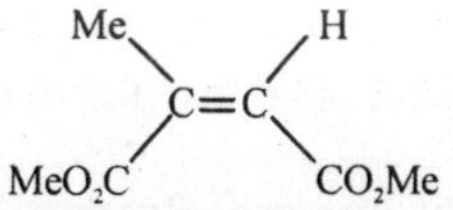

methyl citraconate

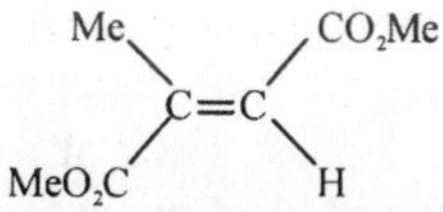

methyl mesaconate

The chemical shift of the olefinic proton in each isomer is different (and so is the methyl proton shift).

(d) *Mass Spectrometry:* In general, trans-isomers give molecular ions of higher intensity than those of the corresponding cis-isomers. Also, the greater the steric effects in the molecule, the greater is this difference in the intensities. Similarly, the intensities of the fragment ions are greater for the trans-isomer than for the cis-, and this difference is increased by using electrons of lower energy.

Method of Surface Films

Long-chain geometrical isomers which contain a terminal group capable of dissolving in a solvent will form surface films, but only the trans-form can form a closepacked film, *e.g,* the long-chain unsaturated fatty acids.

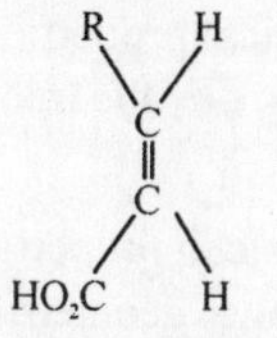

cis- (or Z-) form

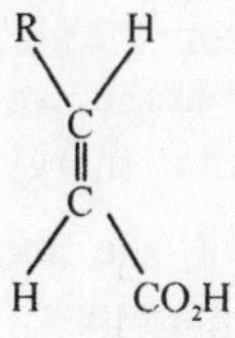

trans- or (E-)

Method of Formation of Solid Solutions

In compounds which owe their property of geometrical isomerism to the presence of an olefinic bond, the shape of the trans-form is similar to that of the, corresponding saturated compound, whereas that of the cis-form is different, *e.g,* the shapes of fumaric and succinic acids are similar, but the shape of maleic acid is different from that of succinic acid. Now, molecules which are approximately of the same size and shape tend to form solid solutions. Thus fumaric acid forma a solid

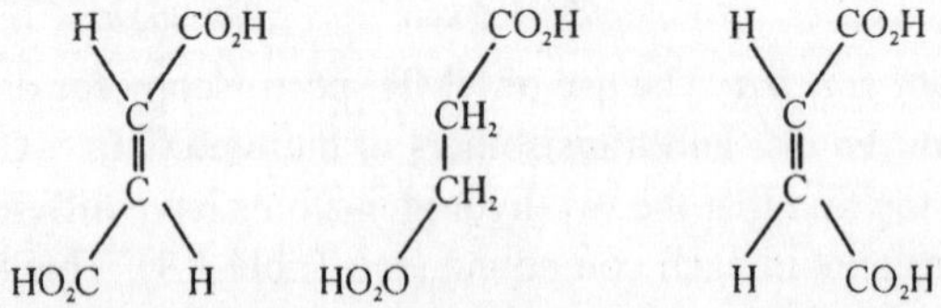

fumaric acid sucinic acid maleic acid

solution with succinic acid, whereas maleic acid does not; hence the configurations of maleic and fumaric acids may be determined.

Methods Based on Generalisations of Physical Properties

Comparison of the physical properties of geometrical isomers of known configurations has led to the following generalisations:

(a) The melting point and intensity of absorption of the cis-isomer are lower than those of the trans.

(b) The boiling point, solubility, heat of combustion, heat of hydrogenation, density, refractive index, dipole moment and dissociation constant (if the compound is an acid) of the cis-isomer are greater than those of the trans.

Based on certain of these generalisations is the Auwers-Skita rule (1915, 1920), viz., in a pair of cis-trans isomers the cis has the higher boiling point, density and refractive index. This rule has been used to elucidate configurations, particularly in terpenoid chemistry, *e.g,* the menthones, but it has now been shown that the use of this rule may give misleading results.

It can be seen from the above physical properties that the trans-form is usually the stabler of the two isomers, *i.e.,* the trans-isomer is the form with the lower internal energy.

Thus, in general, the above physical properties may be used to determine the configurations of unknown geometrical isomers, but the results should always be accepted with reserve, since exceptions are known. Even so, determination of as many as possible of the above physical properties will lead to reliable results, since deviations from the generalisations appear to be manifested in only one or two properties. It should also be noted that where the method of dipole moments can be applied, the results are reliable.

Another method based on generalisations of physical properties is that suggested by Werner. Werner (1904) pointed out that ethylenic cis-trans isomers may be compared with the ortho- and para-isomers in the benzene series, the assumption being made that the melting points of the cis- and ortho-isomers are lower than those of the corresponding trans- and para-isomers, *e.g,*

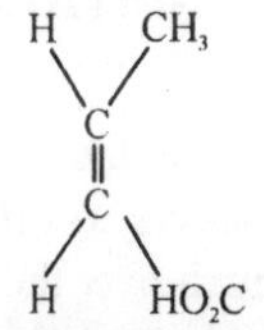

cis- (or Z-) crotonic acid
m.p. 15.5°C

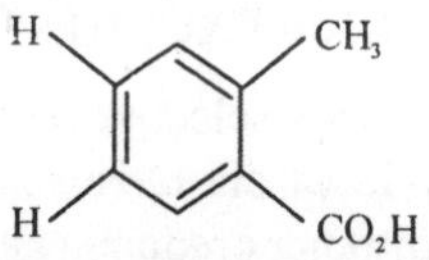

o-toluic acid
m.p. 105°C

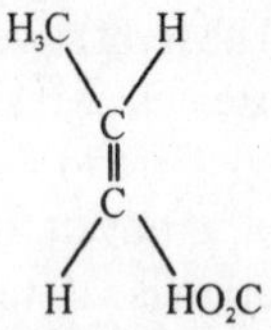

trans- (or E-) crotonic acid
m.p. 72°C

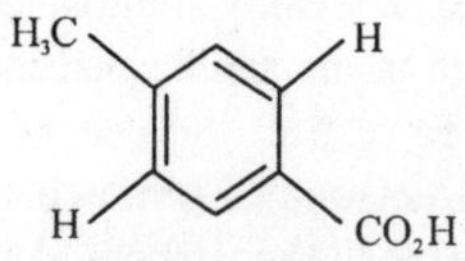

p-toluic acid
m.p. 180°C

Thus, comparison of melting points offers a means of assigning configurations to geometrical isomers. Examination of the above structures shows that, as far as the shape of the molecule is concerned, the benzene ring may be regarded as usurping the function of C=C in the olefinic compound.

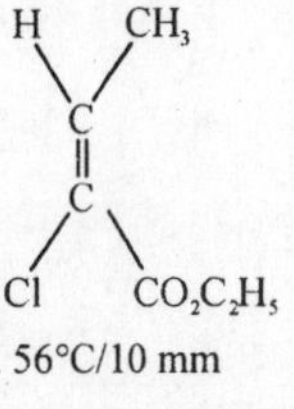

b.p. 56°C/10 mm

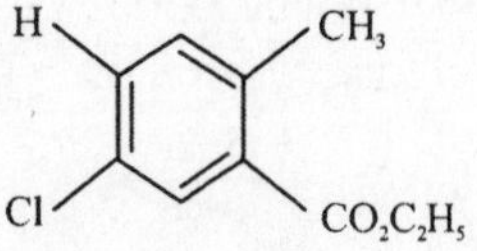

b.p. 122°C

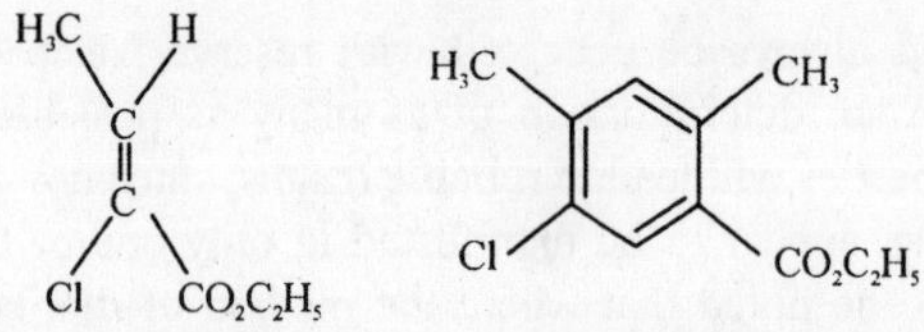

b.p. 61° C/10 m b.p. 130°C

By making use of this idea, it has been possible to assign configurations to difficult cases of geometrical isomerism, *e.g*, there are two ethyl α-chlorocrotonates, and by comparing their physical properties with ethyl 5-chloro-o- and 3-chloro-p-toluates, configurations may be assigned to the chlorocrotonates.

METHOD OF STEREOSELECTIVE ADDITION AND ELIMINATION REACTIONS

The term stereoselective reaction is used in those cases where a given substrate produces diastereoisomeric products in different amounts. If one diastereoisomer predominates very much over the other, the reaction is said to be highly stereoselective.

If the two products are formed in almost equal amounts, the reaction is then said to be weakly stereoselective. The term stereospecific reaction has been used in the same sense as stereoselective reaction, but now the tendency is to restrict the use of stereospecific to a reaction in which different stereoisomers produce different products or act at different rates, *e.g*, the biochemical method of resolution is a stereospecific reaction.

Addition Reactions

(a) *Reduction:* Catalytic hydrogenation of alkenes and alkynes normally gives the cis-addition product.

Thus, the catalytic hydrogenation (Pd) of cis-2,3-diphenylbutene in acetic acid gives almost completely (98 per cent) meso-2,3-diphenylbutane, and the trans-isomer gives the (±)-product (Cardew *et al.,* 1957).

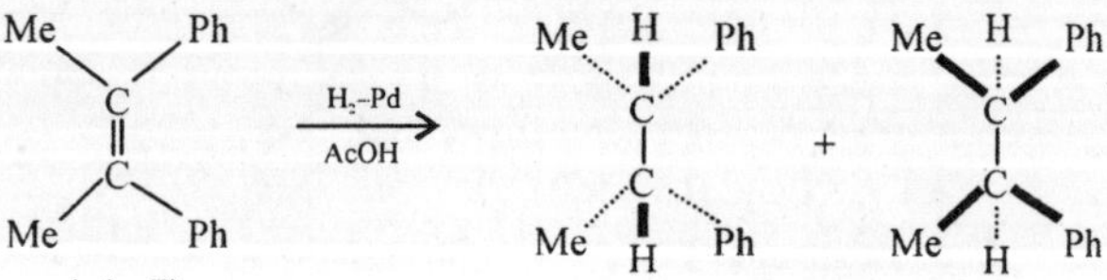

cis (or Z) (both are the same meso-form)

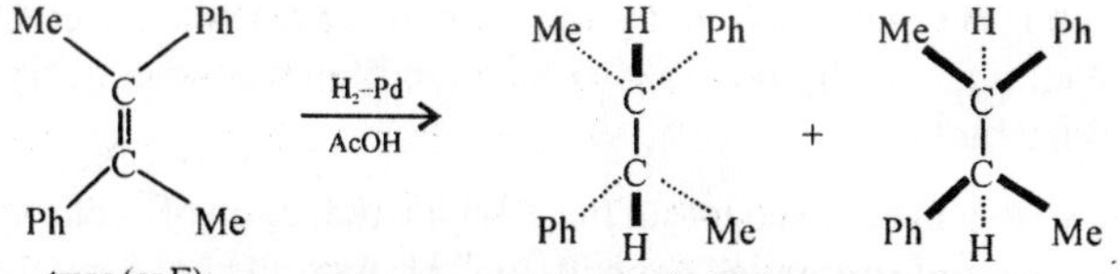

trans (or E) (pair of enantiomers)

The mechanism of catalytic hydrogenation is still not completely understood. It is widely accepted that the hydrogen is adsorbed on the metal surface and is present as hydrogen atoms ($H_2 \rightarrow 2H.$), and that the alkene is also adsorbed on the metal surface. Although the nature of the bonding to the metal in both cases is uncertain, it appears that it is more chemical than physical, and so is described as a chemisorptive bond. Linstead *et al.*, (1942), from their studies on catalytic hydrogenation, proposed that the less hindered side of an unsaturated molecule is adsorbed on the metal surface, and that this is then followed by the simultaneous addition of two hydrogen atoms. In this way, the addition to form the cis-product was explained (attack must be from one side only). Furthermore, since chemisorption has converted hydrogen into hydrogen atoms and has broken one of the multiple bonds, the activation energy of catalytic hydrogenation is considerably lower than that in the uncatalysed reaction.

Later work has shown that this mechanism is unsatisfactory. Isomerisation of alkenes may occur during hydrogenation, *e.g,* Smith *et al.,* (1962) isolated, in addition to n-butane, cis- and transbut-2-ene from the products of the incomplete catalytic hydrogenation of but-l-ene. Also, although cis-addition usually predominates (*i.e.,* the addition is stereselective), trans-addition may occur and even predominate. These results and hydrogen-deuterium exchange experiments lead to the conclusion that addition of the adsorbed hydrogen atoms occurs one at a time and that the reaction is reversible (cf. catalytic dehydrogenation), *e.g,* (asterisks indicate metallic sites):

$$H_2 + CH_2{=}CH_2 \rightleftharpoons \underset{*}{\underset{|}{H}}\quad \underset{*}{\underset{|}{H}} + \underset{*}{\underset{|}{CH_2}}{-}\underset{*}{\underset{|}{CH_2}} \rightleftharpoons$$

$$\underset{*}{\underset{|}{H}} + CH_2{-}\underset{*}{\underset{|}{CH_2}} \rightleftharpoons CH_3{-}CH_3$$

When reducible functional groups are also present, *e.g,* C=O, C≡N, CO_2R, etc., it is usually possible to find conditions to selectively reduce the olefinic bond.

The stereochemical course of the catalytic reduction of cyclic systems–cycloalkenes and aromatics-depends on the nature of the catalyst, *e.g,* the reduction of 1, 2-dimethylcyclohexene in the presence of platinum in acetic acid gives predominantly cis-addition, whereas with palladium trans-addition predominates.

By using suitable conditions, *e.g,* Pd–$BaSO_4$–S as catalyst, it is possible to isolate alkenes when alkynes are reduced.

Now let us consider chemical reduction. The olefinic bond is not reduced by metal and acid, sodium and ethanol, lithium aluminium hydride, etc. unless it is α, β with respect to certain groups, *e.g,* C=O.

Some chemical reagents, however, do reduce olefinic bonds, *e.g,* alkenes are reduced to cis products by di-imide, and this stereospecificity can be explained by the formation of a cyclic T.S.

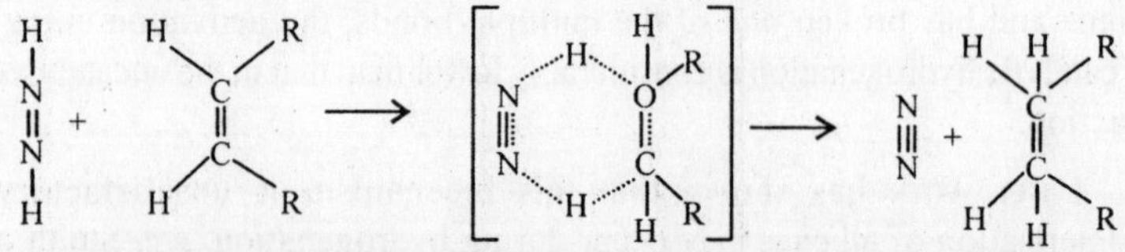

Di-imide is an unstable solid (at low temperature), and so is prepared in situ, *e.g,* by the oxidation hydrazine with hydrogen peroxide, etc. Di-imide is selective in that it reduces carbon–carbon and nitrogen-nitrogen multiple bonds, but does not usually reduce C=O, NO_2, C≡N, etc.

Reduction by di-imide is an example of transfer hydrogenation (hydrogen is supplied by a donor molecule which is itself oxidised). Another donor molecule is cyclohexene (which is oxidised to benzene):

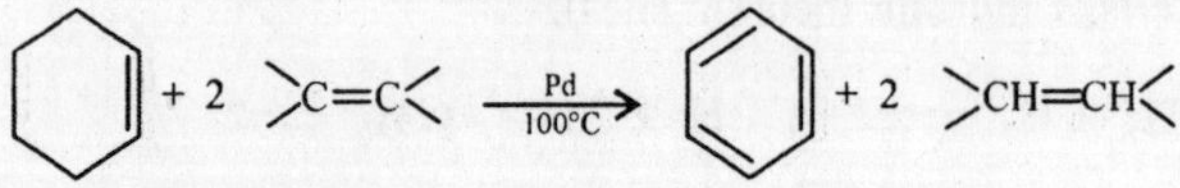

This system reduces many types of functional groups, *e.g,* olefinic and acetylenic bonds, NO_2, –N=N–, etc.

A terminal double bond may be reduced by sodium in liquid ammonia in the presence of an alcohol (MeOH or EtOH; alcohols are stronger acids than ammonia). This method is known as the Birch reduction, and is believed to proceed stepwise via an anionic free radical

$$RCH{=}CH_2 \xrightarrow[e]{Na} RCH\bar{C}H_2 \xrightarrow{EtOH}$$

$$R\dot{C}HCH_3 \xrightarrow[e]{Na} R\ddot{C}HCH_3 \xrightarrow{EtOH} RCH_2CH_3$$

The double bond is also reduced in excellent yield by $NaBH_4$–$PtCl_2$ (Brown *et al.,* 1962).

Alkynes are converted by the Birch reduction into the trans-alkenes.

Hydroboronation (hydroboration): Alkynes are reduced by diborane to trialkenylboranes which, on treatment with propionic acid, give the cis-addition products.

Di-isobutylaluminium hydride also gives cis-addition. The stereospecific cis-addition has been explained by the reaction proceeding stepwise through an intermediate cyclic transition state.

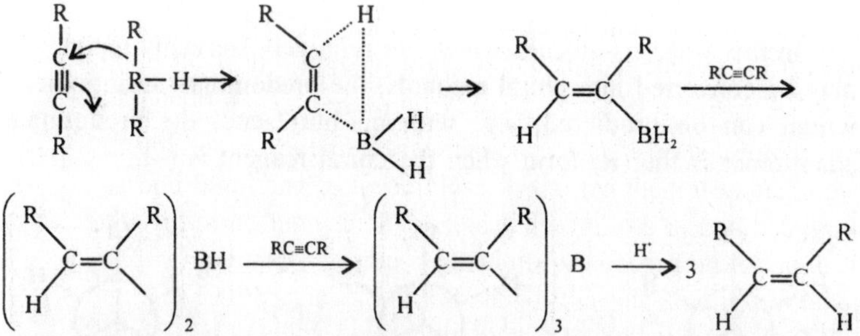

Stereospecific cis-addition also occurs with alkenes to give alkanes and as with alkynes, occurs via a cyclic transition state. This may be formulated as shown:

$$RCH = CH_2 \xrightarrow{BH_3} RCH_2CH_2BH_2 \xrightarrow{RCH=CH_2}$$

$$(RCH_2CH_2{-})_2BH \xrightarrow{RCH=CH_2} (RCH_2CH_2{-})_3B$$

Trialkylboranes are readily oxidised by alkaline hydrogen peroxide to alcohols, the overall hydration being cis. A possible mechanism is one which involves a 1,2-shift.

$$R_3B + \bar{O}OH \rightarrow R_2\overset{R}{B^-} - O - OH \rightarrow R_2B - OR + OH^- \rightarrow$$

$$R_2B - OH + RO^- \xrightarrow{H_2O} R_2BOH + ROH + OH^-$$

Hydroboronation can be carried out with the chiral reagent 'di-3-pinanylborane' (di-isopinocampheylborane), (II), which is prepared from either (+)- or (–)-α-pinene. Also, since oxidation of the (+)-derivative with alkaline hydrogen peroxide gives (–)-isopinocampheol (III) without inversion, the configuration of the borane is established.

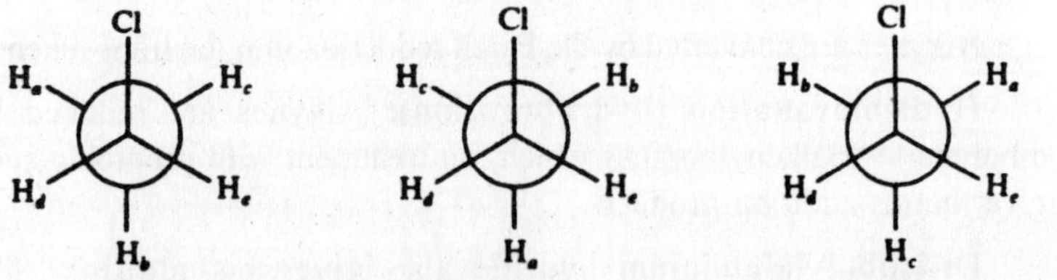

(I) (II) (III)

Fig. 1

In this way, cis-alkenes which are relatively sterically unhindered, may be converted into chiral alcohols, the predominant enantiomer of which can be predicted, *e.g,* with cis-but-2-ene, the predominant enantiomer is the (R)-form when the chiral reagent is (–).

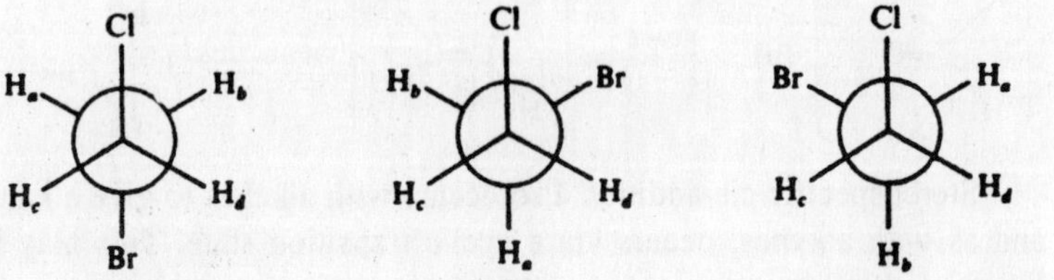

(IV) (V) (VI)

Fig. 2

Asymmetric syntheses with this borane derivative have been used mainly for the preparation of chiral alcohols, but other types of compounds have also been prepared, *e.g,* chiral ketones (by oxidation of the alcohol).

(b) *Hydroxylation:* The configuration of the product formed by hydroxylation of a double bond depends on the nature of the hydroxylating agent used and on the conditions under which the reaction is carried out. Permanganate and osmium tetroxide apparently always give cis-addition, whereas permonosulphuric acid (Caro's acid) and perbenzoic acid give trans-addition.

Table 1

Reagent	*Type of addition*	*Maleic acid*	*Fumaric acid*
$KMnO_4$	cis	mesotartaric acid	DL-tartaric acid
OsO_4	cis	mesotartaric acid	DL-tartaric acid
H_2SO_5	trans	DL-tartaric acid	mesotartaric acid
$C_6H_5COO_2H$	trans	DL-tartaric acid	mesotartaric acid
H_2O_2–O_5O_4	cis	mesotartaric acid	DL-tartaric acid
H_2O_2–SeO_2	trans	DL-tartaric acid	mesotartaric acid

On the other hand, hydroxylation with hydrogen peroxide catalysed by osmium tetroxide in t-butanol gives cisaddition; if the reaction is catalysed by selenium dioxide in t-butanol or in acetone, then the addition is trans. The table above shows the products formed by hydroxylation of maleic and fumaric acids.

With potassium permanganate and osmium tetroxide the cis-addition is readily explained by assuming the formation of a cyclic organometallic intermediate, *e.g*, with permanganate:

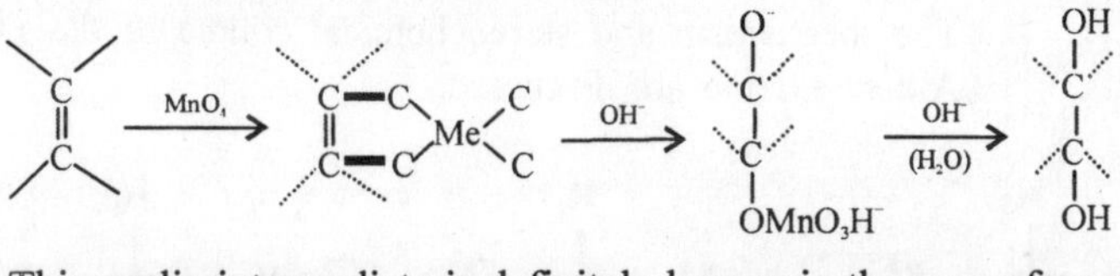

This cyclic intermediate is definitely known in the case of osmium tetroxide for potassium permanganate it may be assumed that the permanganate ion, .MnO_4^-, behaves in a similar manner.

This is supported by the work of Wiberg *et al.,* (1957), who used potassium permanganate labelled with ^{18}O and showed that both glycol oxygen atoms come from the permanganate ion. This also indicates that fission of the cyclic compound occurs between the O and Mn atoms.

With per-acids the hydroxylation results in trans-addition. The first product of oxidation is an epoxide.

Evidence from kinetic studies on solutions of epoxides under high pressure strongly suggests that acid-catalysed hydrolysis is a bimolecular substitution of the conjugate acid (Whalley *et al.,* 1959). This will result in trans-hydroxylation. Thus:

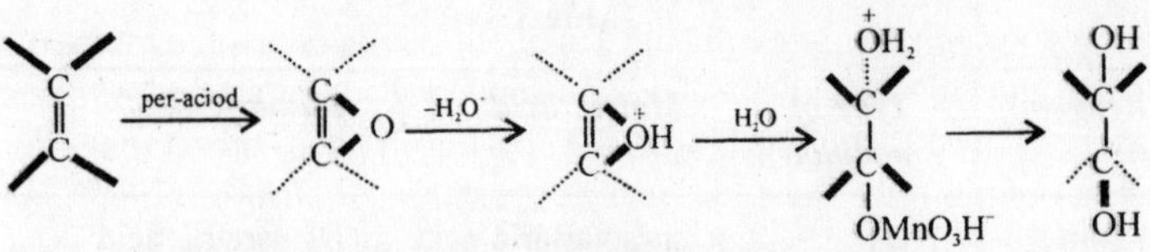

The addition of hydrogen peroxide may result in cis or trans compounds. Which occurs depends on the conditions of the experiment, *e.g,* the catalyst. Where trans-addition occurs, the mechanism may possibly be through the epoxide, but a free hydroxyl radical mechanism could also result in the trans-glycol. Cis-addition in the presence of certain oxides probably occurs via a cyclic intermediate.

(c) (i) *Diels-Alder Reaction:* The addition of a dienophile to a diene in the Diels-Alder reaction is stereospecific; each geometrical isomer forms the cis-addition product. Since it is usually possible to determine the configuration of the cyclic adduct, this offers a means of ascertaining the configuration of the dienophile, *e.g,* butadiene forms adducts with cis- and trans-cinnamic acids, and hence determination of the configurations of the stereoisomeric adducts will determine the configurations of the cinnamic acids; thus: The mechanism and stereochemical course of the Diels-Alder reaction are discussed.

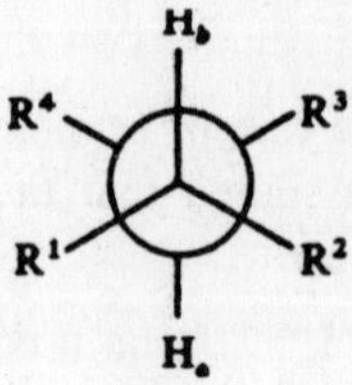

(VII)

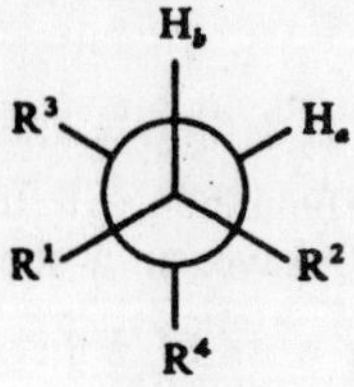

(VIII)

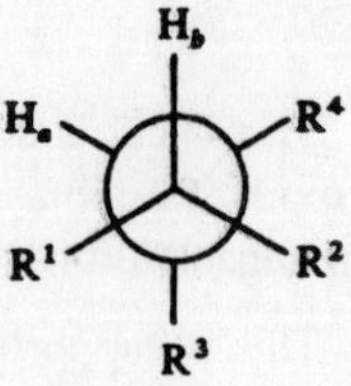

(IX)

Fig. 3

(ii) *Addition of Methylene:* This also is a stereospecific reaction, each geometrical isomer forming the cis-addition product. Thus, *e.g,* Skell *et al.,* (1956, 1959) showed that methylene (carbene), from the photolysis of diazomethane, added to cis- and trans-but-2-ene in a cis-fashion:

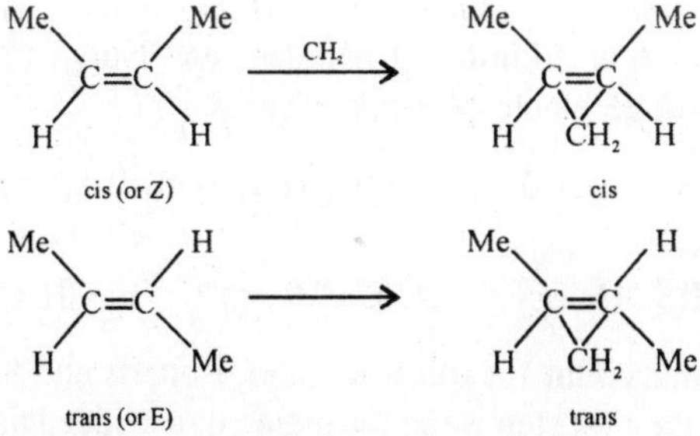

This stereospecificity, however, is lost when the reaction is carried out in the presence of an inert gas (nitrogen), *i.e.,* each substrate now gives a mixture of the cis- and trans-products (Anet *et al.,* 1960). Duncan *et al.,* (1962) also showed that methylene formed by the photolysis of keten added to the above substrates in a non-stereospecific manner.

According to Skell *et al.,* (1956) the stereospecificity of the reaction is due to the addition of singlet methylene, which forms both bonds simultaneously:

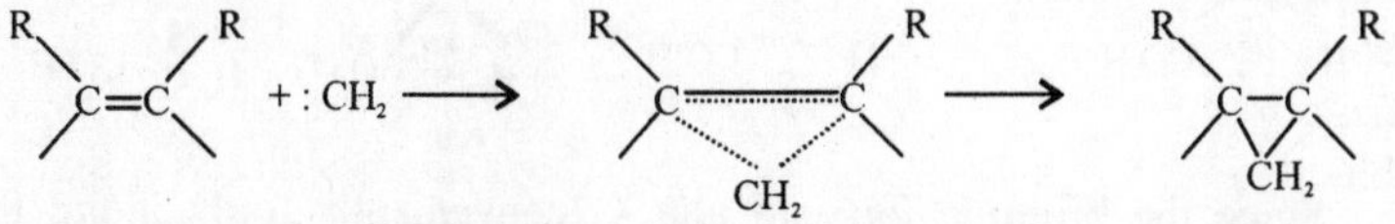

The non- stereo specific reaction k attributed to the addition of methylene in the triplet state:

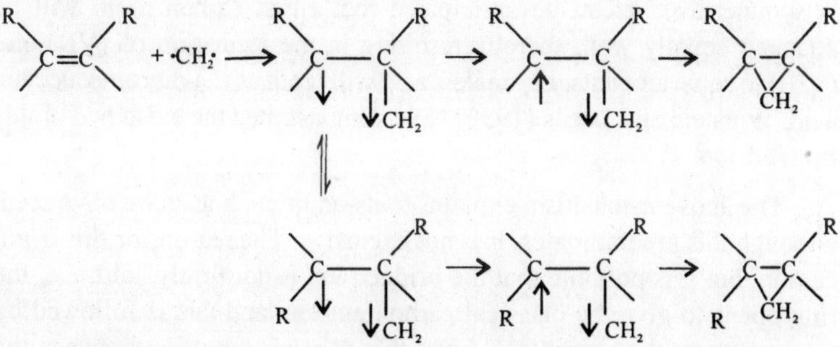

Fig. 4

In the triplet state, the two electrons have parallel spins, and methylene behaves as a diradical. Also, to account for the non-stereospecificity, it is assumed that rotation about the single bond is more rapid than spin inversion.

(d) *Polar Addition of Halogens and Halogen Acids:* All the evidence for the polar addition of halogens and halogen acids indicates a two-stage electrophilic mechanism.

$$CH_2{=}CH_2 \quad Br{-}Br \rightarrow \overset{+}{C}H_2CH_2Br + Br^- \rightarrow CH_2BrCH_2Br$$

$$CH_2{=}CH_2 \quad H{-}Cl \rightarrow \overset{+}{C}H_2CH_3 + Cl^- \rightarrow CH_2ClCH_3$$

In order to account for trans-addition, Roberts and Kimball (1937) suggested that the first step is the formation of a bridged halogenium ion, *e.g,* with bromine the brominium (bromonium) ion is formed first. If a classical carbonium ion were formed first, then one could expect free rotation about the newly-formed single bond and in this case the stereochemical addition would not be the one observed in practice. Thus for maleic acid the reaction may be formulated as follows:

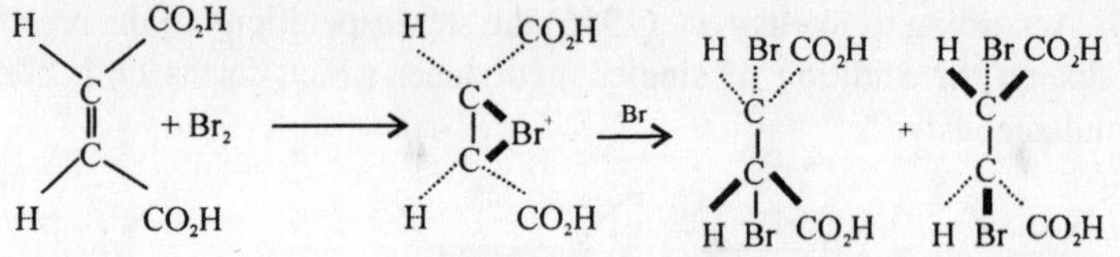

(VII) (VIII)

Since the bromide ion can attack 'conveniently' only along the C–Br$^+$ bonding line and on the side remote from the bromine, a Walden inversion occurs at the carbon atom attacked. Since the brominium ion is symmetrical, it can be anticipated that either carbon atom will be attacked equally well, thereby resulting in the formation of (VII) and (VIII) in equal amounts, *i.e.,* maleic acid will produce (±)-dibromosuccinic acid. Winstein and Lucas (1939) have demonstrated the existence of this bridged ion.

The above mechanism explains trans-addition, but, as we have seen, although this predominates, it is not exclusive. The reason for this is not certain, but it is possible that the bridged ion is not firmly held, *i.e.,* the ring opens to give the classical carbonium ion, and this is followed by rotation about the single C–C bond due to steric repulsion between the carboxyl groups. This would explain the experiments of Michael (1892) that both the maleate ion and fumarate ion add chlorine or bromine to give mainly meso-dihalogenosuccinic acid. The configurations of the products indicate that transaddition has occurred with the fumarate ion but cis- addition with the maleate ion. Roberts and Kimball, however, have explained these results by assuming that the intermediate maleate bro-minium ion (cis) changes to the fumarate brominium ion (trans) due to the powerful repulsions of the negatively charged carboxylate ion groups.

In all of the foregoing examples, addition of halogen and halogen acid has been shown to be predominantly trans, and where the results were not in accord with this, explanations have been offered in terms of steric effects.

This, however, cannot be used to satisfy all cases where 'cis-addition' has occurred, *e.g,* Dewar *et al.,* (1963) found that both cis- and trans-1-phenylpropene add deuterium bromide (or chloride), under conditions giving rise to the polar mechanism, to give predominantly (88 per cent) the cis-addition product.

As an explanation, the authors propose a mechanism involving the rate-determining formation of a classical carbonium ion as an ion-pair (IX) where the halide ion is held

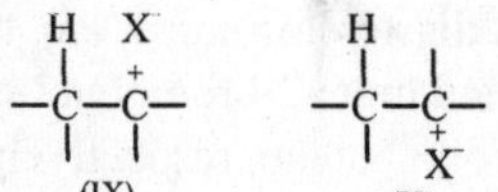

(IX) (X)

on the same side of the original double bond as the entering proton. Collapse of this ionpair thus gives the cis-adduct but, at the same time, rearrangement of (IX) to the isomeric ion-pair, (X), gives the trans-adduct. Dewar suggests that cis-addition should be the rule in electrophilic additions proceeding through a classical carbonium ion, and that the predominance of transaddition, as commonly observed, is due to steric effects. Other examples of cis-addition are also known, *e.g,*

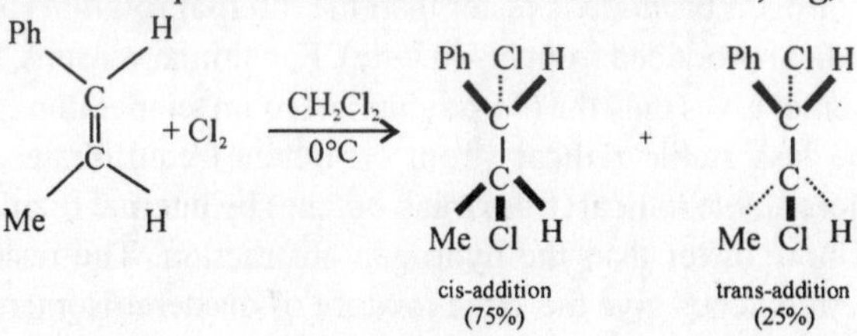

cis-addition (75%) trans-addition (25%)

Fig. 5

On the other hand, trans-butene adds chlorine in the absence of solvent to give exclusively the trans-addition product. It would therefore appear that the stereochemical course of addition of

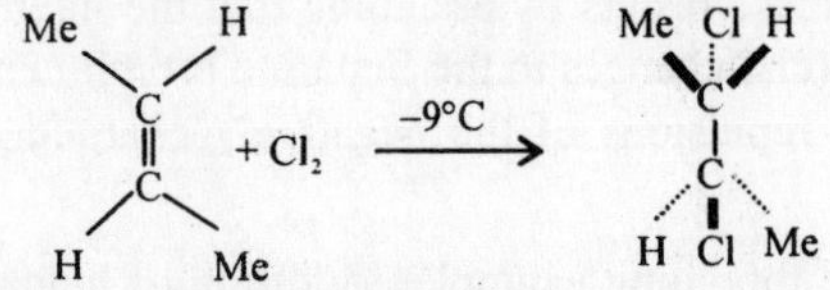

electrophilic reagents depends on the nature of the alkene, the addendum, and on the conditions.

Alkynes undergo nucleophilic addition with strong bases to give predominantly trans-addition, *e.g,*

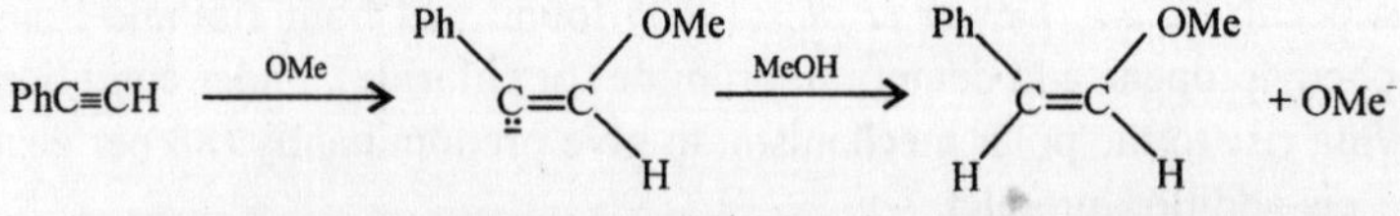

(e) *Free Radical Additions:* The addition of hydrogen bromide to acyclic alkenes in the presence of light or peroxides is a free-radical reaction. The reaction is stereospecifically trans at low temperatures, but at room temperature the two isomers give the same mixture of diastereoisomers, *e.g,* the addition of hydrogen bromide to cis- and trans-2-bromobut2-ene at –78 C gives meso- and (±)-2,3-dibromobutane, respectively. At room temperature, however, both bromobutenes give the same mixture of products (Goering *et al.,* 1959).

The results may be explained as follows. Since attack by the bromine atom can occur equally well at the upper carbon atom from the front (as shown in the equations; note the (E–Z) nomenclature) and the back the cis-isomer gives only the mesoproduct because the abstraction of hydrogen from the hydrogen bromide is faster than the internal rotation (of the free radical produced from cis-butene). For similar reasons, the trans-isomer gives only the (±)-product. At room temperature, however, the less stable radical (from cis-butene) equilibrates with the more stable radical (from trans-butene) by internal rotation which is now faster than the hydrogen abstraction. The result is that both butenes give the same mixture of diastereoisomers. In both cases, the addition-of hydrogen is trans with respect to the bromine atom.

Free radical addition to cyclic alkenes is trans; no rotation is now possible, *e.g,* Goering *et al.,* (1952) showed that 1-bromo-cyclohexene added hydrogen bromide in the presence of benzoyl peroxide to give cis-l,2-dibromocyclohexane (note that the two bromine atoms are equatorial and axial, respectively).

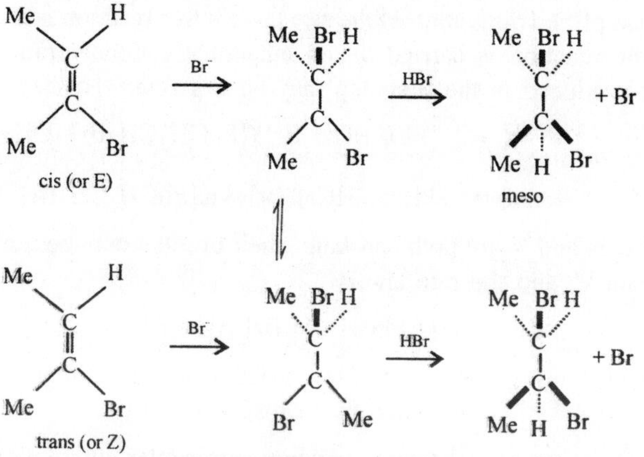

Fig. 6

ELIMINATION REACTIONS

In alkene-forming eliminations, two mechanisms are possible, E_1 and E_2, *i.e.*, unimolecular and bimolecular eliminations.

E1 $\quad H—CR_2—CR_2—Z \underset{}{\overset{slow}{\rightleftharpoons}} Z^- + H—CR_2—\overset{+}{C}R_2 \xrightarrow{fast}$

$H+ + CR_2{=}CR_2$

E2 $\quad Y^-\ H—CR_2—CR_2—Z \rightarrow YH + CR_2 = CR_2 + Z^-$

Two other mechanisms are also believed to operate in certain circumstances. All of these mechanisms of elimination belong to the ionic type.

Evidence for the El mechanism has been obtained in several ways, *e.g*, the rate law for many eliminations is first-order (alkyl halide only), *i.e.*, rate = k[RX]. This is consistent with the mechanism given.

Evidence for the E2 mechanism comes from the fact that the rate law for many eliminations is second-order, *i.e.*, rate = k[RX] [B]. There is, however, a difficulty here.

The second-order rate law is consistent with the one-step mechanism given above, but it is also consistent with the following two-step mechanism

$$Et\bar{O} + PhCH_2CH_{2Br} \overset{fast}{\rightleftharpoons} EtOH + Ph\bar{C}HCH_2Br$$

$$Ph\bar{C}H - CH_2 - Br \xrightarrow{slow} PhCH = CH_2 + Br^-$$

A simplified derivation of the rate law for this reaction is as follows. Since the reaction is carried out in ethanol, the concentration of the ethanol produced in the first step may be neglected. Hence

$$K = [Ph\bar{C}HCH_2Br]/([PhCH_2CH_2Br]\,[OEt^-])$$

$$rate = k[Ph\bar{C}HCH_2Br]=kK[PhCH_2CH_2Br]\,[OEt^-]$$

Since k and K are both constants, their product may be replaced by a constant k', and the rate law is

$$rate = k'[PhCH_2CH_2Br]\,[OEt^-]$$

i.e., the rate law is second-order. Thus, kinetics cannot distinguish between these two possible elimination mechanisms.

In the second mechanism, a slow unimolecular elimination occurs in the conjugate case (cB) of the reactant, and hence this mechanism is called the E1cB or carbanion mechanism.

Since the first step must be reversible (acid-conjugate base equilibrium), if ethanol containing EtOD is used as solvent, it would be expected that the original bromide would incorporate deuterium.

Skell *et al.,* (1945) examined this reaction and found that after the reaction was half-completed, the recovered bromide did not contain deuterium. Hence, this mechanism is not E_1cB, but is E_2.

More O'Ferall (1970) has presented a great deal of evidence to show that in aqueous solution 9-fluorenylmethanol undergoes base-catalysed β-elimination to form dibenzofulvene by an ElcB mechanism (RCH_2 = fluorene):

$$RCH_2CH_2OH + OH^- \rightleftharpoons R\ddot{C}H - CH_2 - OH \rightleftharpoons RCH{=}CH_2 + OH^-$$

Schlosser *et al.,* (1967) have obtained evidence for the existence of the E2cB mechanism in the reaction between cis-styryl chloride and an organolithium base.

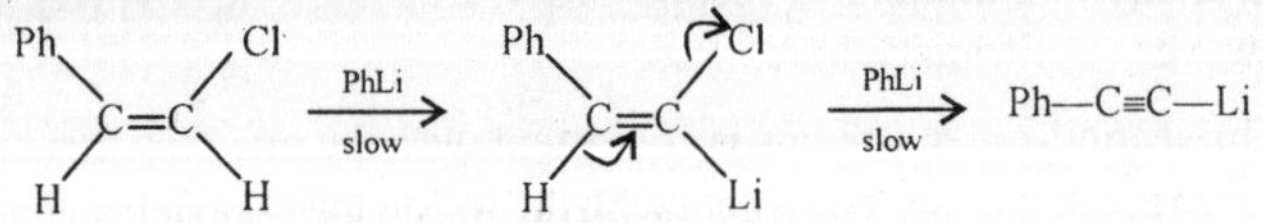

Many examples in the literature show that trans elimination occurs more readily than cis, *e.g*,:

(a) Michael (1895) showed that reaction I was about 50 times as fast as 2.

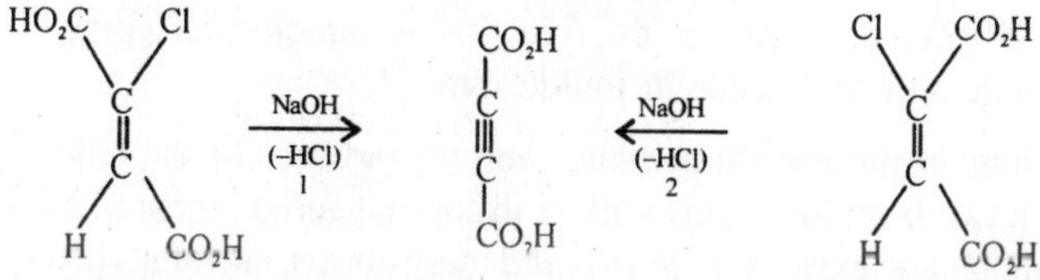

(b) Chavanne (1912) showed that reaction I was about 20 times as fast as 2.

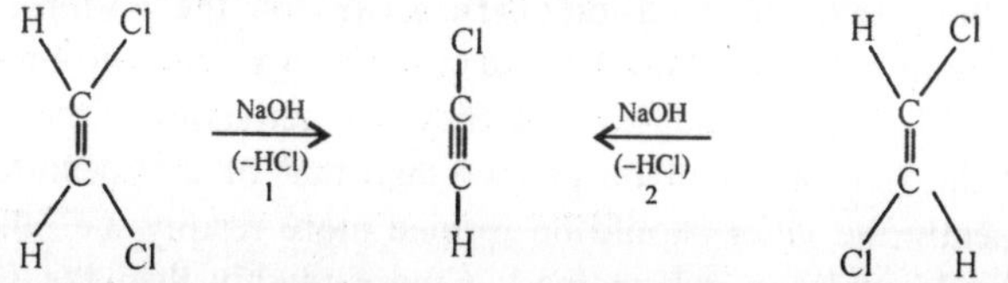

(c) Cristol (1947) showed that the β-isomer of hexachlorocyclohexane under went base-catalysed elimination with great difficulty, whereas under the same conditions all the other known isomers readily underwent second-order elimination to form trichloro benzenes; the β-isomer is the only one in which all the 1,2-HCl pairs are cis. Thus, in the E_2 reaction, the trans requirement is necessary.

According to Hughes and Ingold, bimolecular elimination reactions (E2) take place when the two groups (to be eliminated) are trans and the groups and the two carbon atoms (to which the groups are attached) all lie in one plane. In this way the planar transition state will be readily formed.

As the proton is being removed from the β-carbon atom by the base, the 'liberated' covalent pair of electrons attacks the α-carbon atom from the rear, thereby forming the double bond with displacement of the halogen atom. This type of sequence is not possible when the β-hydrogen atom is cis to the halogen atom, *i.e.*, the stereoelectronic requirement is that the groups which are to be eliminated must be in the trans position.

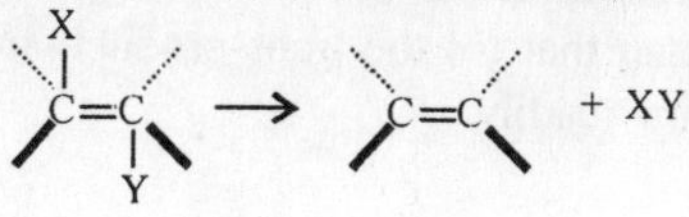

The foregoing evidence for trans-elimination has come from a study of 'rigid' molecules. However, there is also abundant evidence obtained from the study of the products and rates of reaction involving acyclic compounds. A classical example is the debromination of 2,3-dibromobutane by means of potassium iodide in acetone solution. Winstein *et al.,* (1939) showed that this reaction is bimolecular (first order in dibromide and first order in iodide ion).

Thus, in the transition state, the two carbons (of the CBr groups) and the two bromine atoms will be in the staggered position. Now, 2,3-dibromobutane exists in (+)-, (–)- and meso-forms, and it has been shown that the (±)-form gives cis-butene, whereas the meso-form gives trans-butene. If we accept the debromination mechanism proposed by Mulders *et al.,* (1963). In the bridged-ion intermediate from the (±)-form, the two methyl groups become eclipsed; in the meso-form a methyl group becomes eclipsed with a hydrogen. Thus, the energy of activation of the transition state of the (±)-form will be greater than that of the meso-form and consequently the latter should be formed more readily, *i.e.,* the meso-form should undergo debromination more readily than the (±)-form. Winstein *et al.,* (1939) have shown that this is so in practice, the rate of debromination being about twice as fast.

Other examples of E2 reactions which show that trans-elimination occurs readily are the following.

Cram *et al.,* (1952) have shown that the base-catalysed dehydrobromination of the diastereoisomeric 1-bromo-l,2-diphenylpropanes (I and II) gives alkenes that can only arise by trans elimination (I is the erythro compound and II is the threo).

Cram *et al.,* (1956) also examined the elimination reaction of the following onium ion with base (Hofmann exhaustive methylation):

$$\text{PhCHMeCHPhNMe}_3^+\}\ \text{I}^- \xrightarrow{\text{OEt}^-} \text{PhMeC=CHPh}$$

This onium ion exists in two forms, threo and erythro, and the results were that the threo-compound gave the trans-alkene and the erythro-compound the cis-alkene; this is in keeping with trans elimination. The rates of elimination, however, were very different, the threo-form reacting over 50 times as fast as the erythro. In the cis-product, the two phenyl groups become eclipsed and hence the energy of activation for this product is greater than that for the trans-product, and consequently the latter is formed more readily.

A more complicated example of elimination is the case of 2-bromobutane. This, on dehydrobromination, forms trans- and cis-but-2-ene in a ratio of about 6:1 (Lucas *et al.,* 1925). This result may be explained by trans-elimination occurring simultaneously from two different conformations, and one might expect at first sight that, since the staggered conformation is more stable than the skew, the population of the former is greater than that of the latter and consequently the transproduct would predominate.

According to the Curtin-Hammett principle, however, this explanation is incorrect. This principle may be stated as follows: Provided that the activation energy of the reaction is large compared with the barrier to rotation, *i.e.,* the rates of formation of the products are slower than the rate of interconversion of the conformers, the relative amounts of the products do not depend on the relative populations of the conformations, but only on the energies of the transition states leading to the products. However, although the product ratio is independent of the population ratio of the conformers, the rate of formation of the products does depend on the population ratio.

As we have seen above, in the cis-transition state, the two large groups become eclipsed, and consequently the activation energy for this reaction is higher than that which gives the transproduct. Hence the trans-product is formed faster than the cis-. Furthermore, it should also be noted that since both reactions proceed by trans-elimination, both conformations satisfy the stereoelectronic requirements equally well.

The Curtin-Hammett principle should always be considered when attempting to analyse the observed preferred configuration of a product in reactions involving a conformationally mobile system.

An interesting point that now arises is: What is the mechanism when the two eliminated groups cannot assume the trans-position? This was first answered by postulating somewhat complicated mechanisms, but now there is a great deal of evidence to show that bimolecular eliminations may, in certain circumstances, proceed by a cis-elimination (cf. addition reactions, above). Cristol *et al.,* (1960) examined the following elimination reaction, where Y = SMe_2^+ or NMe_3^+ (the latter is the Hofmann exhaustive methylation). The reaction was shown to occur by the E2 mechanism, and so it follows that 1-phenylcyclohexene is formed by a cis-elimination and 3-phenylcyclohexene by the normal trans-elimination. Ingold (1963) has explained the cis-elimination by proposing that the bond changes no longer occur simultaneously (as in trans-elimination). The more difficult it is to detach the C_α group (*i.e.,* Y), the further ahead will the proton transfer proceed on C_β (as compared with the breaking of the C_α bond), thereby building up a negative charge on C_β. As this negative charge builds up, the electron-pair is becoming increasingly available for forming the double bond when C_α is 'free'. This uncoupling of the C_α and C_β bond charges, because of the sequence described, permits phenyl conjugation to control the orientation of the elimination, *i.e.,* cis-elimination is now favoured. Since it has been shown (from other experiments) that it is more difficult to detach the NMe_3^+ group than the SMe_2^+ group, the presence of the former will therefore lead to more ciselimination than the latter.

The cis-eliminations described above occur by the E2 mechanism. Now let us consider the El mechanism. It has long been believed that the product composition (substitution and elimination products) in S_N1 reactions was independent of the leaving group (in halides, p-toluenesulphonates, etc.). The experiments which led to this conclusion were carried out in fairly highly basic solvents, *e.g,* aqueous ethanol. However, Winstein *et al.,* (1963) found that the product composition of the solvolysis of a series of t-butyl compounds in water was independent of the leaving group, but this was no longer true when the solvolysis was carried out in more weakly basic solvents such as ethanol or acetic acid. The explanation offered is that in the latter solvents ionisation to an ion-pair occurs, and the counter-ion then assists in the removal of a proton from the carbonium ion, different counter-ions reacting at different rates. Cram *et al.,* (1963) have also shown that the product composition of a series of 2-phenylbut-2-yl compounds in acetic acid depends on the nature of the leaving group. On the other hand, Skell *et al.,* (1963), by using erythro- and threo-3-deuterobut-2-yl p-toluenesulphonates in different solvents, were able to estimate the amount of cis- and trans-elimination by determining the deuterium content of the but-2-enes produced.

Thus, four products are possible from the erythro-isomer, two types of elimination occurring with each conformation (cf. the Curtin-Hammett principle above). It was shown that cis-elimination varied from 0 per cent with sodium ethoxide in ethanol to 100 per cent when nitrobenzene was the solvent. The explanation offered for the latter result is that, in nitrobenzene, the departing p-toluenesulphonate anion assists in the removal of the β-proton (or deuteron) to yield cis-elimination products.

Molecular Eliminations: Most eliminations occur by a polar mechanism (see above), whereas cyclic eliminations are unimolecular non-polar reactions which take place in one step. Most occur when the compound is subjected to pyrolysis, and proceed via a cyclic T.S. This mechanism is supported by the fact that these reactions show a negative entropy of activation (a cyclic structure has less freedom than an open-chain structure).

In general, trans-elimination occurs in the polar mechanism (see above), whereas in pyrolytic eliminations, the elimination is cis. This is a consequence of the formation of a cyclic T.S.; both the eliminated proton and `leaving group' are in the cis-position.

An advantage of cyclic eliminations is that no carbon skeleton rearrangement occurs (cf. dehydration of alcohols). Also, in the pyrolysis of esters and xanthates of s- and t-alcohols, mixtures of alkenes are produced, but the terminal alkene is favoured over the non-terminal alkene.

INTERCONVERSION OF GEOMETRICAL ISOMERS

The cis-isomer, being usually the more labile form, is readily converted into the trans-form under suitable physical or chemical conditions. The usual chemical reagents used for stereomutation are halogens and nitrous acid, *e.g,*

$$\text{malefic acid} \xrightarrow{Br_2;\, I_2} \text{fumaric acid}$$

$$\text{oleic acid} \xrightarrow{HNO_2} \text{elaidic acid}$$

Other methods such as distillation or prolonged heating above the melting point also usually convert the cis-isomer into the trans, but, in general, the result is a mixture of the two forms.

The conversion of the trans-isomer into the cis may be effected by means of sunlight, but the best method is to use ultraviolet light in the presence of a trace of bromine.

Photochemical cis-trans Isomerisations: cis-trans-Isomerisation can be carried out (usually in solution) by irradiation alone or in the presence of a sensitiser or a catalyst. In general, an equilibrium mixture is reached, the cis-trans ratio remaining constant no matter how much longer the irradiation is continued.

This condition, called a photostationary state, is independent of which isomer is the starting material and always contains predominantly the cis-isomer. The actual ratio of the cis-trans forms, however, depends on a number of factors, *e.g,* solvent, temperature, nature of the sensitiser, etc.

The cis-trans isomerisation of the stilbenes has been examined in great detail, and so we shall use this as our example. The evidence is strongly in favour that the reaction proceeds via a triplet state (Ph–CH–CH–Ph), but since the cis-trans ratio depends on the temperature, several theories have been proposed to explain the isomerisation.

According to one theory, the cis- and trans-forms give different excited states, and the temperature effect is due to the existence of an energy barrier between these excited states. Because of steric effects, the cis-isomer has a higher energy content than the trans in the ground state, and this is also (usually) the case for the corresponding excited states. Also, because of this larger steric effect, the molar absorptivity of the cis-isomer is (usually) lower than that of the trans-isomer.

Hence, the population of the trans excited state is greater than that of the cis. These excited states can return to their respective ground states or, by rotation, the cis* (T1) and trans* (T,) become interconvertible. However, in this interconversion, only the cis* trans* is energetically favourable, but nevertheless, the overall process favours the trans cis interconversion. The reason for this is that the trans population is greater than that of the cis, and the cis → cis* is more difficult than the trans → trans*.

In photosensitised cis-trans isomerisation, the sensitiser excitation energy (of the T state) must be higher than that of the cis- and trans-isomers for energy transfer to occur. Hence, in practice, as long as the sensitiser has a triplet energy state above a certain minimum, the cis-trans ratio is very little affected by the nature of the sensitiser. Thus, if benzophenone is the sensitiser (donor), the mechanisms for the isomerisation of stilbene (acceptor) may be written:

Actually, the mechanism given is an over-simplification; there is evidence to show that another triplet state of stilbene (intermediate in energy content between the cis and trans triplet states) may also be produced. This triplet state, because it can be reached directly from the ground state (normally a forbidden transition) has been called a phantom triplet (Hammond, 1960).

(I) $Ph_2CO(S_0) \xrightarrow{hv} [Ph_2CO]^*(S_1) \rightarrow [Ph2CO]^*(T_1)$

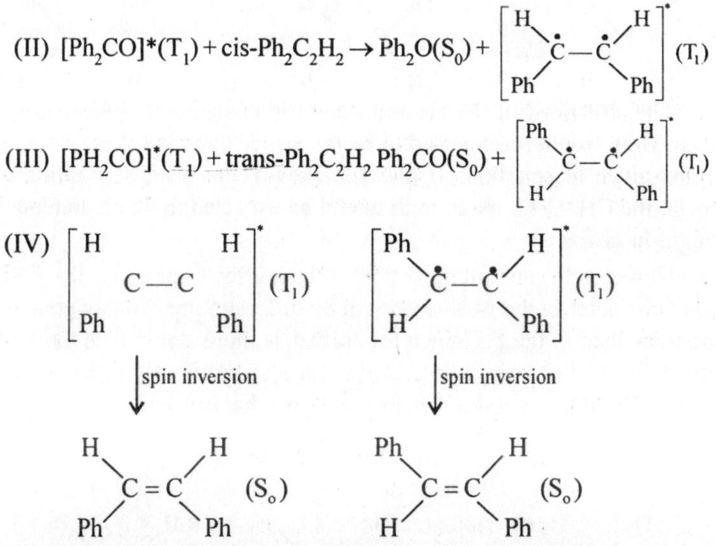

Fig. 7

In addition to photochemical cis-trans isomerisations of alkenes, other systems, *e.g,* oximes, azo-compounds, can be isomerised under similar conditions.

Thermal cis-trans isomerisation is believed to occur by paths similar to those of photochemical isomerisation. Boron trifluoridd also catalyses the conversion of cis- into trans-stilbene. In this case the mechanism is less certain, but a reasonable one is:

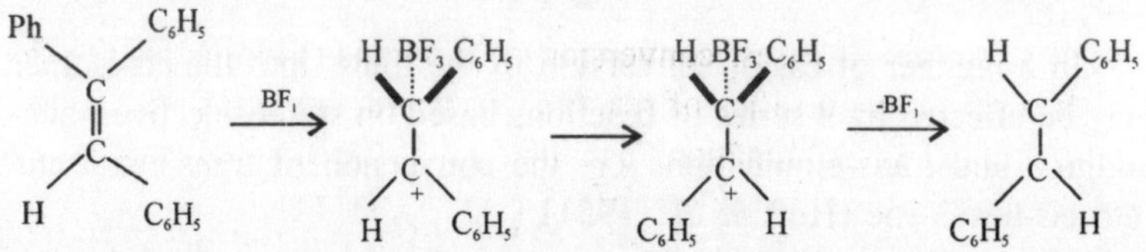

In many compounds containing the group C=C–C=O, the cis-form is readily transformed into the trans in acid solution.

The mechanism of this change is uncertain, but at least one case has been studied in great detail. Noyce *et al.,* (1963), using cis-cinnamic acid as substrate, showed that the rate-determining step is the addition of a proton, and the mechanism proposed is:

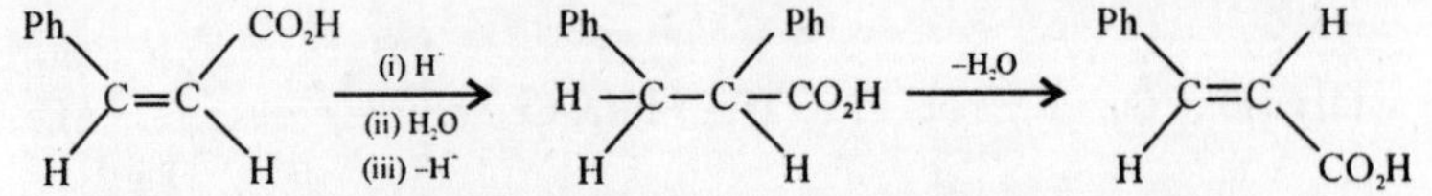

The structures of the cis and trans triplet states of stilbene are not clear. Both would be expected to be resonance hybrids (of the canonical forms given in equations II and III, above), but because rotation can occur, the CH–CH bond in both would be expected to be predominantly single in character.

However, because of the different steric effects operating, the double bond character in the two forms will be different, presumably greater in the trans than in the cis, since the former is more stable than the latter. Photochemical cis-trans isomerisation can also be effected in the presence of, *e.g*, bromine or iodine. A probable mechanism is:

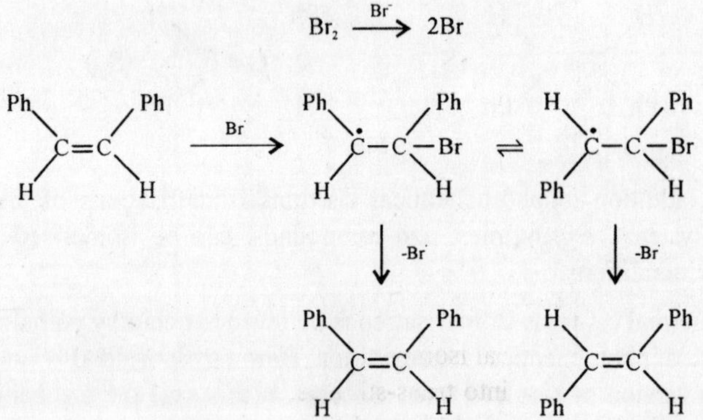

Fig. 8

In a number of cases, conversion of the trans- into the cis-isomer may be effected by a series of reactions based on stereoselective trans-addition and trans-elimination, *e.g*, the conversion of trans-hex-3-ene into cis-hex-3-ene (Hoff *et al.*, 1951).

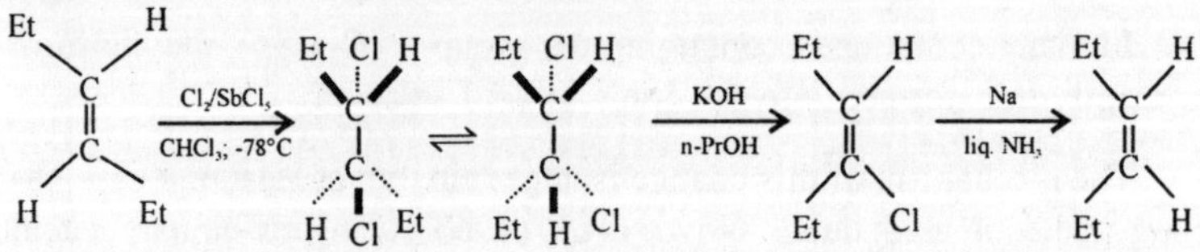

Fig. 9

CYCLIC COMPOUNDS

Geometrical and optical isomerism may exist in any sized ring. In the following account, the saturated rings are treated as rigid flat structures, and the groups attached to the ring-carbon atoms are regarded as being above or below the plane of the ring. Furthermore, the examples described deal only with those cases in which the chiral centres are part of the saturated ring system. In general, the pattern of optical isomerism followed by cyclic compounds is similar to that of the acyclic compounds. The main difference between the two is that, since there is no free rotation about ring-carbon atoms, geometrical isomerism may therefore be manifested as well as optical isomerism. On the other hand, geometrical isomerism may exist without optical isomerism. Since no ambiguity arises with the cis-trans nomenclature of geometrical isomerism of cyclic structures, this nomenclature has been retained only for cyclic systems.

When we come to describe cyclohexane we shall also introduce the principles of conformational analysis of ring systems, and then apply these principles to the various ring systems.

Classification of Monocyclic Systems: Monocyclic systems have been classified according to the number of carbon atoms in the ring: small rings, 3-4; common rings, 5-7; medium rings, 8-11; large rings, 12–. As we shall see, many chemical properties depend on the class of the cycloalkane, and these differences in behaviour have been explained largely in terms of steric strain.

CYCLOPROPANE TYPES

Molecule (I) contains one chiral centre (*), and is not superimposable on its mirror image molecule (II). Thus (I) and (II) are enantiomers, *i.e.*, a cyclopropane derivative containing one chiral centre can exist in two

(i)

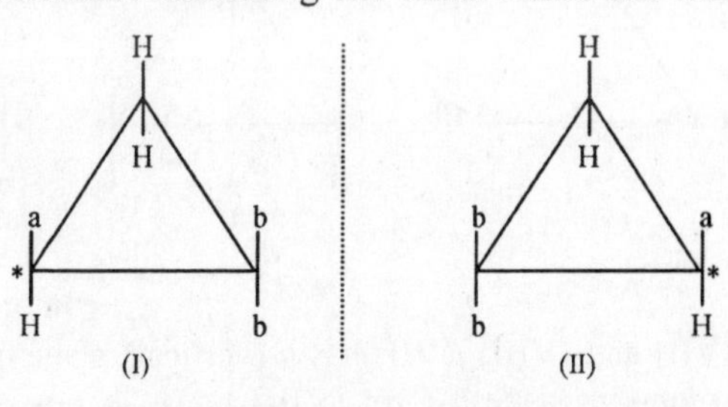

(I) (II)

Fig. 10

optically active forms. Molecule (III) contains two different chiral centres, and since it has no elements of symmetry it is not superimposable on its mirror image molecule. Thus (III) can exist in two optically active forms (and one racemic modification). Structure (III), however, is capable of exhibiting geometrical isomerism, the two geometrical isomers being (III) and (IV).

(ii)

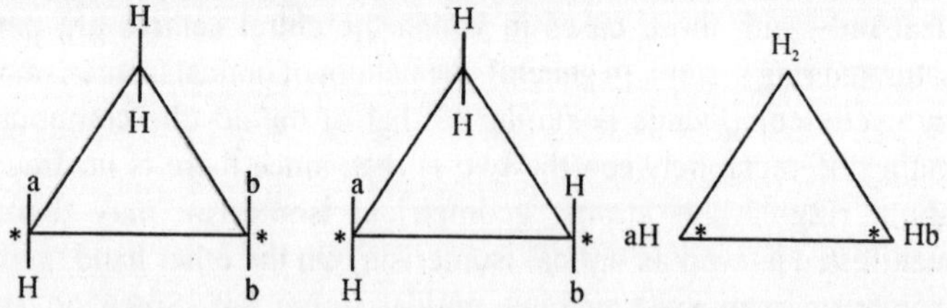

(III) (IV) (V)

Fig. 11

Now (IV) also contains two different chiral centres, and these are not disposed towards each other as in (III). Since (IV) possesses no elements of symmetry, it can also exist in two optically active forms which are different from those of (III).

Thus (V), which may be regarded as the non-committal way of writing the configurations (III) and (IV), is similar, as far as optical isomerism is concerned, to the acyclic molecule CabdCabe, *i.e.,* there are four optically active forms in all (two pairs of enantiomers).

In general, any monocyclic system can exist in 2^n optically active forms, where n is the number of different chiral centres. Molecule (VI) contains two similar chiral centres and can exist as geometrical

(iii)

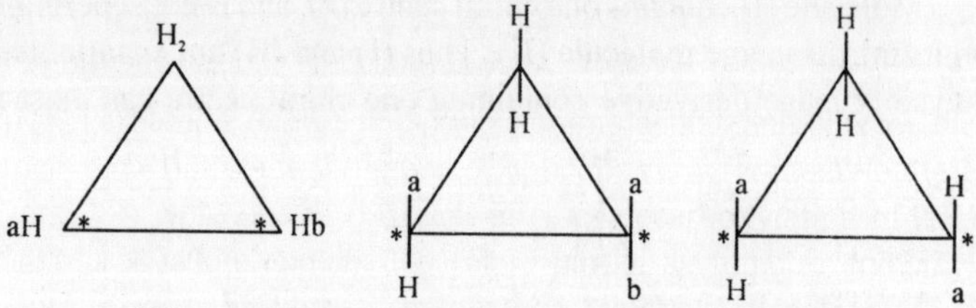

(VI) (VII) (VIII)

Fig. 12

isomers (VII) and (VIII). (VII) has a (vertical) plane of symmetry and therefore represents a meso-form. (VIII), however, possesses no elements of symmetry and can therefore exist in two optically active forms (and one racemic modification). (IX) contains three different chiral centres and can therefore exist in 23 = 8 optically active forms (four pairs of enantiomers). Each pair of enantiomers is derived from the four geometrical isomers (X–XIII). Inspection of these configurations shows that all of them possess no elements of symmetry. (XIV) contains two similar asymmetric carbon atoms, and the third carbon atom is pseudo-asymmetric.

(iv)

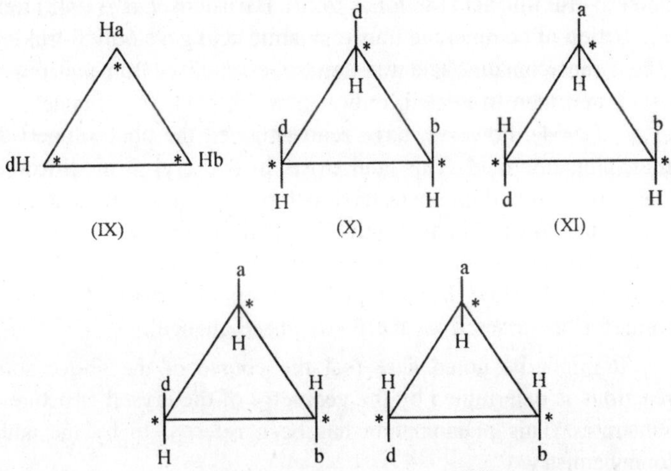

(IX) (X) (XI) (XII) (XIII)

Fig. 13

(v)

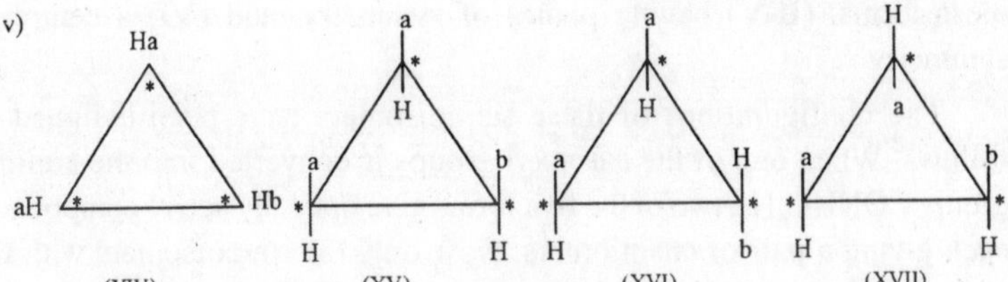

(XIV) (XV) (XVI) (XVII)

Fig. 14

Three geometrical isomers, (XV)–(XVII), are possible; (XV) and (XVI) each possess a (vertical) plane of symmetry, and therefore each represents a meso-form. (XVII), however, possesses no elements of symmetry and so can exist in two optically active forms (and one racemic modification). (XVIII) contains three similar asymmetric carbon atoms which are all pseudo-asymmetric. Two geometrical isomers are possible, (XIX) and (XX), both of which possess at least one (vertical) plane of symmetry, and therefore represent meso-forms

CYCLOBUTANE TYPES

Two examples of the cycobutane type are truxillic and truxinic acids; truxillic acid is 2,4-diphenylcyclobutane-1,3-dicarboxylic acid, and truxinic acid is 3,4-diphenylcyclobutane-1,2-dicarboxylic acid cis-Cinnamic acid (allocinnamic acid), on irradiation with light, forms mainly β-truxinic acid and trans-cinnamic acid, together with some of the dimer of the latter, α-truxillic acid (de Jong, 1929). Bernstein *et al.,* (1943) found that irradiation of commercial trans-cinnamic acid gave only β-truxinic acid. When trans-cinnamic acid was slowly recrystallised from aqueous ethanol, dried, and then irradiated, only α-truxillic acid was obtained. Schmidt *et al.,* (1964), however, have reinvestigated the photo-dimerisation of trans-cinnamic acid. This acid exists in two crystal modifications, the stable α-form and the metastable β-form. It was shown that the α-form gives pure α-truxillic acid only. The β-form gives pure β-truxinic acid at temperatures where the β→α phase transformation of the monomer is sufficiently slow, but at higher temperatures α-truxillic acid is also formed (this arises from the β→α phase change).

It might be noted here that the course of the above solid-state reactions is determined by the geometry of the crystal structures of the substrates. This phenomenon has been referred to by the authors as topochemistry.

Truxillic Acid: This acid can exist theoretically in five stereoisomeric forms, all of which are known (the acid is of the type 1). All five are meso-forms, (II-V) having planes of symmetry, and (VI) a centre of symmetry.

The configurations of these stereoisomers have been assigned as follows. When one of the carboxyl groups is converted into the anilido-group, $CONHC_6H_5$, two of the five forms give optically active compounds, each giving a pair of enantiomers. Now only the stereoisomers with the two phenyl groups in the trans-position can produce chiral molecules under these conditions; the remaining forms will each have a (vertical) plane of symmetry. Thus only (IV) and (VI) satisfy the necessary conditions. One of these is known as the α-acid (m.p. 274°c) and the other the γ-acid (m.p. 288°C). This then raises the problem: Which is which? This is readily answered by the fact that of the anilido-derivatives of these two acids, only one can be dehydrated to a cyclic N-phenyl imide, $-CO-N(C_6H_5)-CO-$. This reaction can be expected to take place only when the two carboxyl groups are in the cis-position. Therefore (IV) is γ-truxillic acid, and (VI) is α-truxillic acid (since the acid with the melting point 288°C has been called the γ-acid). By considering the ease of formation of the cyclic anhydride, the configurations of the remaining three stereoisomers may be determined. Two form anhydrides readily, and therefore one of these acids must be (II) and the other (III). The third acid does not form its own anhydride, but gives a mixture of the anhydrides produced by (II) and (III). Thus the third acid, epi-truxillic acid, is (V). The final problem is to decide which of the two, (II) and (III), is peri-truxillic acid, and which is e-truxillic acid peri-Truxillic acid, under the influence of aluminium chloride, undergoes an internal Friedel-Crafts reaction to form a truxonic acid (VII) and a truxone (VIII).

This is only possible when the phenyl and carboxyl groups are in the cis-position. Thus (II) isperi-truxillic acid, and therefore (III) is e-truxillic acid.

Truxinie Acid: This acid can exist theoretically in six geometrical isomeric forms, four of which are resolvable; thus ten forms in all are possible theoretically. Truxinic acid is of the type (IX) and the six geometrical isomers possible are (X–XV). (X) and (XI) are meso-forms (each has a plane of symmetry); (XII–XV) are resolvable (theoretically), since all possess no elements of symmetry. The configurations of these stereoisomers have been determined by methods similar to those used for the truxillic acids; it appears, however, that only four of these six forms are known with certainty, viz., β, δ, ζ and neo.

CYCLOPENTANE TYPES

A number of examples involving the stereochemistry of the five-membered ring occur in natural products, *e.g,* camphoric acid, furanose sugars. In this section we shall discuss the case of 2,5-dimethyl-cyclopentane-1, 1-dicarboxylic acid. This acid can exist in two geometrical isomeric forms, which may be differentiated by decarboxylation, the cis-isomer giving two monocarboxylic acids, (I) and (II), and the trans-isomer one monocarboxylic acid, (III). All three acids contain two similar asymmetric carbon atoms and one pseudo-asymmetric carbon atom. Both (I) and (II) possess a (vertical) plane of symmetry, and are therefore meso-forms; (III) possesses no elements of symmetry, and can therefore exist in two optically active forms (and one racemic modification). All the possible forms are known, and (I) and (II) have been differentiated as follows. The diethyl ester of the cis-dicarboxylic acid (IV) can be partially hydrolysed to the monoethyl ester, which most probably has the configuration (V). This is based on the assumption that the carbethoxyl group on the same side as the two methyl groups is far more resistant to attack than the other carbethoxyl group because of the steric effect. Decarboxylation of (V) gives (VI), and this, on hydrolysis, gives (I). Thus the configuration of (I) (and therefore also of (II)) is determined. These assignments are supported by the fact that (II) is esterified more rapidly than (I). Also, (I) can be isomerised to (II) on heating in acetic acid containing hydrogen chloride. This indicates that the latter is the more stable isomer, *i.e.,* is the trans-isomer.

CYCLOHEXANE TYPES

The stereochemistry of cyclohexane and its derivatives presents a detailed example of the principles of conformational analysis. The principles are the same as those for acyclic compounds, but because of the `rigidity' of cyclic systems, additional problems are involved. On the basis of the tetrahedral theory, two forms are possible for cyclohexane, neither of which is planar. These two forms, known as boat and chair conformations were first proposed by Sachse who also pointed out that both are strainless.

The chair form is 'rigid' (in the sense that it resists distortion), and when it is transformed into the boat form some angular deformation is necessary. The energy barrier in this process has been determined from NMR spectral data; it is about 37.7–46.0 kJ mol^{-1} (Sheppard *et al.,* 1961; Jensen *et al.,* 1962. This value is large enough for each conformation to retain its identity, but is not large enough to prevent their rapid interconversion at room temperature. Thus it is not possible to isolate each conformation.

The chair and boat forms are both free from angle strain, but because of differences in steric strain and bond opposition strain, the two forms differ in energy content. According to Hassel *et al.,* (1963), there is a small amount of angle strain, the ring angles being ~111.5° (not the normal angle of 109.5°). Figures 1.50(a) and 1.50(b) represent the chair and boat conformations and the directions of the C–H bonds. In the chair conformation, all the C–H bonds on adjacent carbons are in the skew position. On the other hand, in the boat conformation there are four skew interactions (1,2; 3,4; 4,5 and 6,1) and two eclipsed interactions (2,3 and 5,6). At the same time, there will also be some bond opposition strain for these two pairs of eclipsed bonds, and also steric repulsion between the two flag-pole (fp) hydrogens (at 1 and 4), which are 1.83 A apart (see Table below). Hence, the total strain in the boat conformation is larger than that in the chair conformation, and consequently the former is less stable than the latter. The boat form, however, is flexible and can readily be distorted into many shapes, and in these the hydrogen eclipsings and the flag-pole interactions are reduced. According to Hendrickson (1961), the twist-boat contains 6.7 kJ mol^{-1} less energy than the classical boat form. Several workers have calculated the energy difference between the flexible and chair forms of cyclohexane, *e.g,* Johnson *et al.,* (1960), from measurements of heat of combustion and other measured quantities, have found that the energy difference is 22.2 ± 1.26 kJ mol^{-1} (at 25°C; vapour phase). This value has been confirmed by the work of Allinger *et al.,* (1960); their value is 24.7 ± 2.5 kJ mol^{-1}. The flexible boat form of cyclohexane is about one to two in a thousand at 25°C. Hassel *et al.,* (1943) were the first to show, by means of electron diffraction studies, that at room temperature the cyclohexane molecules are mainly in the chair conformation. Also, the examination of cyclohexane derivatives by X-ray and electron diffraction has shown the presence of the chair conformation (see also below). However, the boat conformation has been found in certain molecules, but their number is relatively small.

The nature of the intermediate in the transformation of one chair form into the other is not certain. Jensen *et al.,* (1962) believe that the transition state (of the intermediate) is the structure, approximately halfway between the chair and the twist-boat form.

The twelve hydrogen atoms in the chair conformation are not equivalent; there are two sets of six. In one of these sets the six C–H bonds are parallel to the threefold axis of symmetry of the molecule; these are the axial (a) bonds (they have also been named ε- or polar bonds). In the other set the six C–H bonds make an angle of 109° 28′ with the axis of the ring (or ± 19° 28′ with the horizontal plane of the ring); these are the equatorial (e) bonds (they have also been named κ-bonds). On the other hand, it can be seen that the 'end' of the boat is different stereochemically from the chair conformation; the various C–H bonds have been named: flag-pole (fp), bowsprit (bs), boat-equatorial (be), and boat-axial (ba).

Angyal and Mills (1952) have calculated the distances between the various hydrogen atoms (and carbon atoms) in both the chair and boat conformations. Since the boat conformation occurs in relatively few cases, in the following account we shall study mainly the problem of the chair conformation. Inspection of the above table shows that for hydrogen atoms, the interactions 1e, 2e, 1e, 2a, and 1a, 3a are about the same.

Furthermore, a study of accurate scale models has shown that with any axial substituent (which is necessarily larger than hydrogen), the 1a, 3a interactions are larger than 1e, 2e or 1e, 2a interactions. Using these principles, we can now proceed to study the conformations of cyclohexane derivatives.

Because of the mobility of the chair conformation, one chair form is readily converted into the other chair form, and in doing so all a- and e-bonds in the first now become e- and a-bonds, respectively, in the second.

Both forms are identical and so cannot be distinguished. If, however, one hydrogen is replaced by some other atom or group, the two forms are no longer identical, *e.g,* methylcyclohexane. In the a-methyl conformation there are 1,3-interactions acting, whereas in the e-methyl conformation these interactions are absent; instead, the weaker 1,2-interactions are acting. Thus the energy content of axial conformation is greater than that of the equatorial, and consequently the latter will be the preferred form. Hassel (1947) has shown experimentally from electron-diffraction studies that the e-methyl conformation predominates in methylcyclohexane. Hassel *et al.,* (1950) have also shown that in chlorocyclohexane the e-form also predominates and that very little of the a-form is present. NMR spectroscopy has also been used as a method for determining conformation. The NMR spectra of unsubstituted cycloalkanes (except cyclopropane) show a τ-value around 8.5 p.p.m. due to the methylene protons (cf. acyclic methylene value, 8.7). As we have seen, cyclohexane exists in two stable chair conformations (I) and (11) which are undergoing rapid interconversion at room temperature. Let us suppose that cyclohexane is a rigid molecule, *i.e.,* no interconversion occurs.

Then, because the environments of axial and equatorial hydrogen are different, the chemical shifts will be different. The general rule is that axial protons absorb upfield with respect to equatorial hydrogen. At the same time, coupling occurs between axial and equatorial hydrogen attached to the same carbon atom, and coupling also occurs between vicinal hydrogen. Since the coupling constant depends on the dihedral angle, J for vicinal axial-equatorial hydrogen (skew; dihedral angle of 60°) will be smaller (2–4 Hz) than that (5–12 Hz) for vicinal axial axial hydrogen (traps; dihedral angle of 180°). The overall result would be an NMR spectrum in which multiplets would have different J values. At room temperature, cyclohexane is not a rigid molecule, and since the interconversion rate is rapid, both axial and equatorial hydrogen have the same average environment, thereby giving rise to a single sharp line. When the spectrum of cyclohexane is measured at about −100°C, two signals are observed, one due to axial and the other to equatorial hydrogen. Under these conditions, the interconversion rate is sufficiently slowed down for each type of proton to show its own chemical shift.

Now let us consider a monosubstituted cylohexane, *e.g*, chlorocyclohexane (III and IV). Since the equatorial chlorine conformation (III) is favoured, the populations of (III) and (IV) are different, and consequently the time spent by a given hydrogen atom (axial or equatorial) will be longer in conformation (III) than in (IV). Even so, because of rapid interconversion at room temperature, only one broad proton signal is observed. When the temperature is cooled below −55°C, two broad peaks are observed.

A detailed investigation of the infrared specta of cyclohexanes and particularly of rigid systems (such as steroids) has shown that the absorption maximum of a given substituent group depends on whether its orientation is axial or equatorial, *e.g*,

a-Cl, ~690; e-Cl, ~740 cm^{-1}

a-OH, 1000–1010; e-OH, 1030–1040 cm^{-1}.

In general, the stretching frequency (C–Z) for an equatorial orientation is higher than that for an axial one by about 10–50 cm^{-1}. This applies to the C–O stretching frequency in cyclohexanols (see above), but not to the O–H bond. An a-axial bromine atom has very little effect on the C=O stretching frequency, but an α-equatorial bromine atom increases the C=O stretching frequency by about 20 cm^{-1}. These generalisations have been used to assign orientations to substituents.

Measurements of ORD and CD curves are very valuable for assigning conformations to cyclohexanones. Mass spectrometry may also be used to distinguish between cis and trans isomers of cyclohexane derivatives.

DISCUSS THE CONFORMATIONS OF DISUBSTITUTED CYCLOHEXANES

Here we have a number of factors to consider: position isomerism, stereoisomerism (geometrical and optical), the relative sizes of the two substituents, and the nature of the substituents.

(i) *1,2-Compounds:* It should be noted that in these cis-compounds one substituent must be axial and the other equatorial. If the substituents differ in size, the 1,3-interactions will be most powerful when the larger group is axial. Thus the conformation with the lower energy will be the one in which the larger group is equatorial, *i.e.,* this is the preferred form. An example of this type is cis-2-methylcyclohexanol; the methyl group is larger than the hydroxyl, and so the preferred form can be expected to be la-hydroxyl,2e-methyl. This has been shown to be so in practice. In general, the greater the difference in size between the two substituents, the greater will be the predominance of the form with the larger group in the equatorial conformation. The classical formula of the cis-compound when the two substituents are identical has a plane of symmetry and is therefore not resolvable. If the conformational diagram is inspected, then for the cis-compound with $Y_1 = Y_2$, there are no elements of symmetry and hence it is not superimposable on its mirror image.

When this conformation flips, the resulting conformation is the mirror image of the original one (this is clearly seen from the equivalent conformation). Since the strain is identical in the original and flipped conformations, their populations are equal and consequently this 1,2-disubstituted cyclohexane ($Y_1 = Y_2$) is optically inactive by external compensation. However, this type of compound has never yet been resolved, and this is due to the fact that the two enantiomers are readily interconvertible. In effect, each enantiomer undergoes very rapid autoracemisation. Whether Y_1 and Y_2 are identical or not, the two conformations are different, and because of the 1,3-interactions the ex-form will be the preferred form. Furthermore, this form will be more stable than the cis-isomer (a, e-form). An example that illustrates this is 2-methylcyclohexanol.

The transform has been shown to be more stable than the cis; the latter is readily converted into the former when heated with sodium, and also the reduction of 2-methylcyclohexanone (with sodium and ethanol) produces the trans-alcohol. Both the classical formula and the e,e- (and a,a) conformation of the trans-1,2-compound (whether Y_1 and Y_2 are identical or not) are not superimposable o.. their mirror images. Also, neither conformation can be converted into its mirror image by flipping. Hence, a trans-l,2disubstituted cyclohexane exists as a pair of enantiomers (whether $Y_1 = Y_2$ or not).

(ii) *1,3-Compounds:* The two trans-conformations are identical when the two Y groups are identical. The cis-e,e-form will be more stable than the cis-a,a, and will also be more stable than the trans-e,a-conformation, *e.g,* the most stable conformation of 1,3-dimethylcyclohexane has been shown to be the cis-l,3-e,e-form. It should be noted that this situation is the reverse of that of the 1,2-d imethylcyclohexanes. The Auwers-Skita rule has been shown to break down when applied to 1,3-disubstituted cyclohexanes: the reverse holds good. Allinger (1954) modified the rule for cyclohexanes as follows: The isomer which has the higher boiling point, refractive index and density is the one with the less stable configuration. Thus, according to this rule, the trans- l,3-disubstituted cyclohexanes have the higher physical constants (the trans-fdrm has more axial substituents than the more stable cis-form); *e.g,* Macbeth *et al.,* (1954) have shown that the physical constants of (±)-trans-3-methylcyclohexylamine are higher than those of its cis-isomer.

(iii) *1,4-Compounds:* The two cis-conformations are identical when the Y groups are identical. Also, the trans-e,e-form will be more stable than the cis-a,e-form.

The arguments used for the disubstituted cyclohexanes can also be applied to the higher substituted cyclohexanes. As the result of a large amount of work, the following generalisations may be made:

(i) In cyclohexane systems, mono-, di-, tri- and poly-substituted derivatives always tend to take up the chair conformation whenever possible.

(ii) The chair conformation with the maximum number of equatorial substituents will be the preferred conformation. This generalisation, however, is only satisfactory when the internal forces due to dipole interactions or hydrogen bonding are absent. When these are present, it is necessary to determine which forces predominate before a conformation can be assigned to the molecule.

(iii) The energy barriers between the various conformations are too small to prevent interconversion.

Now let us apply these generalisations to various compounds. Cyclohexane-l,3-diol has been shown to have the diaxial rather than the diequatorial orientation. This can beexplained on the basis that intramolecular hydrogen bonding (which has been shown to be present from infrared spectra) can stabilise the diaxial but not the diequatorial form. The conformation of the ring is the chair form, but when intramolecular hydrogen bonding is possible between groups in the 1 and 4 positions, the molecule may then assume a boat conformation rather than the chair in which this hydrogen bonding is not possible. However, what have been considered to be boat conformations may be twist-boat forms in some cases, *e.g,* Stolow (1961) showed, from infrared spectral data, that 2,5-di-t-butylcyclohexane-l,4-diol exists in several conformations, with one having the twist-boat form. Another class of compounds which are of great interest are the derivatives of cyclohexanone. If we assume that the C atom of the CO group is trigonally hybridised, then this has little effect on the shape of the ring. On the other hand, if we assume tetrahedral hybridisation with banana bonds, then very little effect on the shape of the ring would be expected.

Also, since the oxygen atom in acetaldehyde is eclipsed with a hydrogen atom in the preferred conformation, it follows that cyclohexanone has no appreciable strain. Because the axial hydrogen is absent at the carbonyl group, an axial group at C-3 has only one 1,3-interaction (at C-5), and similarly an axial group at C-5 has only one 1,3-interaction (at C-3).

Hence, an axial group at C-3 or C-5 in cyclohexanone will be more stable than in cyclohexane. This has been referred to as the '3-alkyl ketone effect'. First, we shall consider 2-bromocyclohexanone; the two possible chair forms are:

On the basis that a substituent preferably takes up an equatorial conformation, it would therefore be expected that the conformation 2e-bromocyclohexanone would be favoured. Infrared studies, however, have shown that the a-bromo conformation predominates. This has been explained as follows. The C–Br and C=O bonds are both strongly polar, and when the bromine is equatorial the dipolar repulsion is a maximum, and a minimum when the bromine is axial. Since the axial form predominates, this equatorial dipolar repulsion must therefore be larger than the 1,3-interactions. When, however, other substituents are present, the 1,3-interactions may become so large as to outweigh the dipolar effect and the bromine would now be equatorial. Such is the case with 2-bromo-4,4dimethylcyclohexanone. Now let us consider cyclohexane-1,4-dione. Le Fevre *et al.,* (1935) showed that this compound has a dipole moment (μ) of 1.2 D. Since the chair form has $\mu = 0$, the conformation must be some other one. Mossel *et al.,* (1963) have examined this dione by dipole measurement and X-ray analysis and conclude from their results that the molecule has a twist-boat conformation. The foregoing discussion has been confined to determining preferred conformations. If we now examine the stereochemistry of cyclohexane derivatives, we find that, up to the present time, the number of geometrical (and optical) isomers obtained from a given cyclohexane derivative is in agreement with the number that can be expected from a planar ring with the substituents lying above and below the plane of the ring. We shall now, therefore, discuss the stereochemistry of some cyclohexane derivatives from the, classical point of view.

(i) *Hexahydrophthalic Acids (cyclohexane-1,2-dicarboxylic acids):* Two geometrical isomers are theoretically possible, the cis, (I), and the trans, (II). Molecule (I) has a plane of symmetry, and therefore represents the meso-form; (II) has no elements of symmetry, and can therefore exist in two optically active forms (and one racemic modification). All of these possible forms are known, and it has been found that the cis-compound, (I), forms a cyclic anhydride readily, whereas the transcompound, (II), forms a cyclic anhydride with difficulty.

(ii) *Hexahydroisophthalic Acids (cyclohexane-l,3-dicarboxylic acids):* Two geometrical isomers are possible; the cis-form (III) has a plane of symmetry, and therefore represents the meso-form; (IV) has no elements of symmetry, and can therefore exist in two optically active forms (and one racemic modification). All of these forms are known; the cis-isomer forms a cyclic anhydride, whereas the trans-isomer does not.

(iii) *Hexahydroterephthalic Acids (cyclohexane-l,4-dicarboxylic acids):* Two geometrical isomers are possible; the cis-form (V) has a plane of symmetry, and the transform (VI) a centre of symmetry. Hence neither is optically active. They may be distinguished by the fact that the cis-isomer forms a cyclic anhydride, whereas the trans-isomer does not. The cyclic anhydride will have a boat c nformation.

(iv) *Inositol (hexahydroxycyclohexane):* There are eight geometrical isomers possible theoretically, and only one of these is not superimposable on its mirror image molecule; thus there are nine forms in all (and also one racemic modification). If we imagine that we looking down at the molecule, and insert the groups which appear above the plane of the ring, then the eight geometrical isomers may be represented as follows: Examination of these configurations shows that all except one–the one labelled resolvable–have at least one plane of symmetry, and so are all meso-forms. All the meso-forms and both of the optically active forms are known; of these myo-inositol, scyllitol and (+)- and (–)-inositol occur naturally.

(v) *Benzenehexachloride (hexachlorocyclohexane):* Here again eight geometrical isomers are possible theoretically; seven are known, α, β, γ, δ, η, θ; the γ-isomer is a powerful insecticide. All have been shown to exist in the chair form, and the conformations that have been assigned are:

α-, aaeeee; α-, eeeeee; γ–, aaaeee; δ-, aeeeee; ε-, aeeaee.

Of these forms, it is the β- which loses hydrogen chloride with the greatest difficulty. All of the other stereoisomers possess at least one pair of chlorine atoms cis to each other (thus having H and Cl trans). Cristol (1949) has also identified the or-isomer as the (±)-form.

Fused Systems

(a) *Decalins and Decalols:* As we have seen, the boat and chair forms of cyclohexane are readily interconvertible, and the result is that cyclohexane behaves as if it were planar. Mohr (1918), however, elaborated Sachse's theory, and predicted that the fusion of two cyclohexane rings, *e.g,* as in decalin, should produce the cis- and trans-forms which would be sufficiently stable to retain their identities. This prediction has now been confirmed experimentally.

Several conventions have been introduced to rėpresent these isomers. One convention uses, full lines to represent groups above the plane of the molecule, and broken lines to represent those below the plane thus cis-decalin will be (I) and trans-decalin (II). This convention appears to be the one most widely used, but there is another, introduced by Linstead (1937), which is favoured by many.

According to this convention, a hydrogen atom is represented as being above the plane of the ring when drawn as in (III), and below the plane when drawn as in (IV); thus cis-decalin will be (V), and trans-decalin (VI).

The configurations of the decalins are complicated, the complication arising from the fact that a number of strainless modifications are possible, which differ in the type of 'locking', *i.e.,* whether axial or equatorial bonds are used to fuse the rings. According to Hassel *et al.,* (1946), cis- and transdecalins are as shown in Fig. 1.9; the cis-form is produced by joining one axial and one equatorial bond of each ring, whereas the trans-form is produced by joining the two rings by equatorial bonds only; in both cases the cyclohexane rings are all chair forms (see also below). However, Geneste *et al.,* (1964) have proposed a 'doubly-crossed' conformation for cis-decalin. This has about the same energy content as the cis-chair/chair conformation. Johnson (1953) has calculated the difference in energy content between these two forms in the following simple manner.

The trans-form is arbitrarily assigned a value of zero energy, and when this form is compared with the cis, it will be found that the latter has three extra skew interactions involving the two axial bonds (this is shown in the following diagram; the cis-form has 3 staggered and 15 skew arrangements, and the trans-form 6 staggered and 12 skew). On the basis that skew interaction of the hydrogens in n-butane is 3.35 kJ (Pitzer, 1940), the total energy difference between the cis- and trans-forms is 3 $\times$ 3.35 = 1005 kJ. This value agrees well with that of Rossini *et al.,* (1960), from measurements of heat of combustion. It might be noted, in passing, that if these two decalins are regarded as 1,2-disubstituted cyclohexanes, then the trans-form (e,e) would be expected to be more stable than the cis- (e,a). We shall now deal with the determination of configuration in the decalin series. The configurations may be ascertained by using the Auwers-Skita rule. Huckel (1923, 1925), however, isolated two forms of 2-decalol and determined their configurations by the following chemical methods.

2-Naphthol, on hydrogenation in the presence of nickel as catalyst, gave two 2-decalols, (VII) and (VIII), each of which, on oxidation with chromic acid, gave a decal-2-one (IX and X). These two decalones each gave, on oxidation with permanganate, a cyclohexane-1,2-diacetic acid. These diacetic acids were geometrical isomers; one was resolvable and therefore must be the trans-isomer (XII); and the other, which was not resolvable, must therefore be the cis-isomer (XI) (this is the meso-form). Thus the configurations of the two decalols and the two decalones are established:

In addition to the two cyclohexane-1,2-diacetic acids (which are formed by scission of the 2,3-bond of the decalone), two other geometrical isomers were also obtained, viz. cis- and trans-cyclohexane1-carboxyl-2-propionic acids (XIII) and (XIV) (these are formed by scission of the 1,2-bond of the decalone). The conversion of 2-naphthol into two decalols does not prove that the two decalols are the cis and trans-isomers described above.

It is possible that both compounds could have been the cis- and trans-forms of a given decalol; since the carbon atom of the CHOH group in the 2-decalol is asymmetric, it can exist in two configurations, *i.e.,* each decalol, (VII) and (VIII), can exist in two forms; (VIIa) and (VIIIa).

Had the two decalols been the two forms of either (VII) or (VIII), then on their oxidation, only one decalone would have been produced. Since, however, two decalones were obtained, the two decalols must be of the types (VII) and (VIII)-one of each, or even a mixture of the pairs; further proof of the existence of the types (VII) and (VIII) lies in the fact that the two decalones gave geometrical isomers of cyclohexane-1,2-diacetic acid. Consideration of formulae (VIIa) and (VIIIa) shows the presence of three chiral centres in each of the four possible forms, and since all four possess no elements of symmetry, four pairs of enantiomers should be possible theoretically. Actually all eight forms have been isolated, but their configurations have not yet been established with certainty.

There are only two geometrical isomers possible for the decalins, and their configurations have been established by the reduction of the two decalones, (IX) and (X), by means of the Wolff Kishner method each decalone gives the corresponding decalin. It is interesting to note in this connection that Willstatter *et al.,* (1924) found that hydrogenation of naphthalene in the presence of platinum black as catalyst gives mainly cis-decalin, whereas in the presence of nickel as catalyst the main product is trans-decalin. The conﬁgurations of the decalins have also been determined by means of their NMR spectra (see also below).

Various other fused ring systems have also been shown to exhibit the same type of geometrical isomerism as the decalins, *e.g,* the hydrindanols exist in cis- and trans-forms (Huckel *et al.*, 1926), and also the decahydroquinolines and decahydroisoquinolines (Helfer *et al.*, 1923, 1926). It has already been pointed out that in monosubstituted cyclohexanes, the preferred conformation is the one with the substituent equatorial, but owing to the low energy barrier between this and the axial form, the two are readily interconvertible.

In the case of the monosubstituted decalins, the problem is more complicated. In cis-decalin, since ring fusion involves equatorial and axial bonds, the molecule is mobile and can interchange with the other cis-form, *i.e.*, there are two cis-forms possible (XV and XVI), and these are identical and in equilibrium (cf. cyclohexane). This has been shown to be so by Hassel (1950); thus:

Musher *et al.*, (1958) distinguished between cis- and trans-decalin by means of their NMR spectra. cis-Decalin shows one proton peak, whereas trans-decalin shows two. The former isomer is capable of rapid interconversion, but the latter is not since it is a rigid system, and so equatorial and axial protons are held in different environments, the result being two different chemical shifts (cf. cyclohexane, 11).

Now let us consider cis-2-decalol. Here there are four possible conformations which, in pairs, are in equilibrium. Two arise from (XV) (XVa and XVb), and two from (XVI) (XVIa and XVIb).

In (XVa) and (XVIb) the hydroxyl group is equatorial, and so these two conformations contain about the same energy. In (XVIa) and (XVb) the hydroxyl group is axial, and on the basis that an equatorial conformation is more stable than an axial, then (XVa) and (XVIb) will contribute more to the actual state of the molecule than will (XVIa) and (XVb), *i.e.*, the hydroxyl group in cis-2decalol should possess more equatorial character than axial. It is also interesting to note that the two axial forms do not contain the same energy.

In (XVb) the a-hydroxyl group is involved in the normal 1,3-hydrogen interactions (at 4 and 9), but in (XVIa) the interaction is the normal 1,3- with the hydrogen at 4 and the larger 1,3-interaction with the CH, group at 8. Thus (XVIa) should be less stable than (XVb).

In trans-decalin there is only one stable conformation, since the ring fusions use equatorial bonds. If the molecular conformations were 'inverted', the two ring fusions would now have to be axial, and this type of fusion is impossible. The axial bonds on adjacent carbon atoms are pointing in opposite directions and the carbon atoms are too far apart to form a bond in a six-membered ring. It is possible, however, for larger rings to have this conformation. Thus, in trans-2-decalol, there are only two conformations possible, (XVII) and (XVIII). Furthermore, the latter, with the equatorial hydroxyl conformation, would be expected to be more stable than the former (with the axial hydroxyl).

(b) *Polycyclic Systems:* Many natural products contain polycyclic systems consisting of fused six-membered rings, but sometimes they also contain a five-membered ring, *e.g,* steroids. One type of compound which has been studied in great detail is perhydrophenanthrene. Ten stereoisomeric forms are possible: four pairs of enantiomers and two meso-compounds. These are as shown, the method of naming being as follows.

The prefixes cis and trans denote the stereochemistry of the ring fusions, *e.g,* cis-hydrogens at 4b (13) and 8a (14) give cis A/B, trans-hydrogens at 4b (13) and 8a (14) give trans A/B, and cis-hydrogens at 4a (12) and I Oa (11) give cis B/C. On the other hand, the prefixes syn and anti denote the terminal orientation of the rings with respect to each other, *e.g,* if the bond joining rings A and C, *i.e.,* bond 4a-4b (12-13) has cis-hydrogens at these carbon atoms, then rings A and C are syn with respect to each other, and if the hydrogens are trans, rings A and C are anti.

The axial or equatorial orientation of the ring-fusion bonds with respect to the central ring, B, are as shown. In 4 and 5, the rings A/B and B/C are both cis as in cis-decalin, and so these molecules are mobile systems, *i.e.,* they exist in two interconvertible forms. Molecules 1, 2 and 3, however, have at least one trans A/B or B/C, and hence are not mobile systems, *e.g,* for 2 to change into its other form, trans B/C would have to become cis B/C, but this is not possible since in the latter case the ring fusion would have to be a,a (cf. trans-decalin). Also, in isomers 1-5 all three rings are in chair forms, but in 6 the central ring, B, is in the classical boat conformation. This isomer will therefore be the highest energy form of all and so is the least stable. Furthermore, since in 1 all the rings are fused by equatorial bonds, this isomer will be the most stable form; it may be regarded as the equivalent of two trans-decalins. Isomers 2 and 3 are therefore each the equivalent of a cis- and a trans-decalin, and so would be expected to be of equal energy content, but greater than that of 1. Isomers 4 and 5 are each the equivalent of two cis-decalins. It has been found, however, that the interactions in a polycyclic system containing 1,3-axial fusion bonds are greater than those in a similar system containing 1,2-axial bonds. Thus 4 is more stable than 5. The foregoing account has been qualitative, but its basis is the work of Johnson (1953) who estimated the differences in energy between the various isomers. Furthermore, Linstead *et al.,* (1950) have elucidated the configurations of a number of perhydrophenanthrene derivatives, and the work on the stabilities of their compounds is in good agreement with the estimated stabilities. There are five stereoisomeric perhydroanthracenes, and these have the configurations shown.

EFFECT OF CONFORMATION ON THE COURSE AND RATE OF REACTIONS

Since the environments of axial and equatorial groups are different, it may be expected that the reactivity of a given group will depend on whether it is axial or equatorial. Now S_N2 reactions always occur with inversion. Hence if the geometry of the molecule is such as to hinder the approach of the attacking group (Z) along the bonding line remote from the group to be expelled (Y), then the S_N2 reaction will be slowed down. Examination of formulae (I) and (II) shows that the transition state for an S_N2 reaction is more readily formed when Y is axial (I) than when it is equatorial (II). In (I), the approach of Z is unhindered, but in (II) the approach of Z is hindered by the rest of the ring. Thus S_N2 reactions take place more readily with an axial substituent than with an equatorial.

The study of S_N1 reactions in cyclohexane derivatives is made difficult because of the ease with which elimination reactions usually occur at the same time. It can be expected, however, that an SN1 reaction will be sterically accelerated for an axial substituent, since the formation of a carbonium ion will relieve the steric strain due to 1,3-interactions. On the other hand, Since these 1,3-interactions are absent for an equatorial substituent, no steric acceleration will operate in this conformation.

A particularly interesting example of concurrent S_N1 and S_N2 mechanisms and the involvement of neighbouring group participation is the acetolysis of trans-4-methoxycyclohexyl tosylate. Noyce *et al.,* (1960) have shown that the products of the acetolysis of the tosylate labelled with tritium in position 1 (*i.e.,* trans-4-methoxy[l–^{3}H]cyclohexyl tosylate) areas shown in the Chart. This mechanism was proposed to account for the fact that the trans-isomer (non-tritiated) underwent acetolysis about 5.6 times as fast as the cis-isomer (hence n.g.p. in the former), the scrambling of the tritium (hence the intermediate bicyclic oxonium cation), and the fact that the trans-4-methoxy[1–^{3}H]cyclohexyl acetate was produced in larger amount than the corresponding trans-[4–^{3}H]-compound (hence all of the trans-acetate cannot have been formed from the oxonium cation, in which case the tritium should have been scrambled equally between the 1- and 4-positions). So far, we have examined reactions in which the bond linking an atom to the ring is broken. Now let us consider reactions in which this does not occur, *e.g,* esterification and ester hydrolysis by the $A_{AC}2$ mechanism. For the present purpose, we may write the equation as shown:

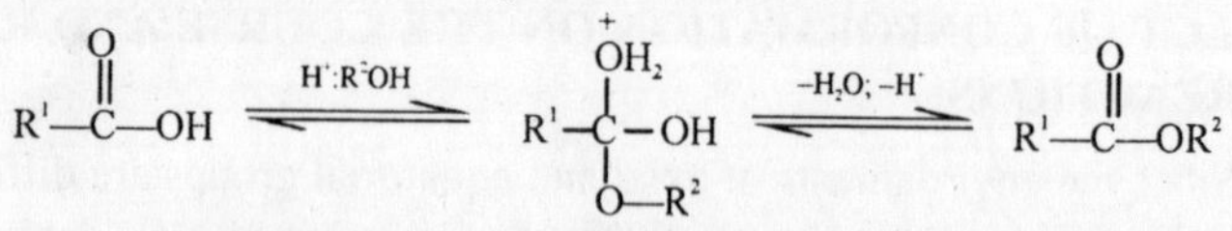

(A)

In the esterification of cyclohexanol ($R^2 = C_6H_{11}$; $R^1 = R$), the reaction proceeds through the intermediate (A). Replacement of the hydroxylic hydrogen in R^2OH by the large group (shown in A) very much increases the non-bonded interactions (in R^2O), far more so for an axial than for an equatorial OH. Hence the former would be expected to be esterified more slowly than the latter. Since hydrolysis by the $A_{AC}2$ mechanism is the reverse of esterification, hydrolysis also proceeds through the intermediate (A).

In this case, the ester already has the large group R^1CO attached to the oxygen atom of R^2O, but in the formation of (A) hybridisation of the carbonyl carbon atom has changed from sp^2 (flat) tp sp^3 (tetrahedral). Thus, the volume of the group has increased in (A) and consequently non-bonded interactions are increased. Hence the axial ester would be expected to undergo hydrolysis more slowly than the equatorial ester.

If we now consider esterification of cyclohexanecarboxylic acid ($R^1 = C_6H_{11}$; $R^2 = R$) and the hydrolysis of its esters, then, since both proceed through the intermediate (A), and using the same arguments as before, the axial conformer would be expected to undergo esterification and hydrolysis more slowly than the equatorial conformer. Similar arguments applied to hydrolysis by the $B_{AC}2$ mechanism will also show that the axial ester would be expected to undergo hydrolysis more slowly than the equatorial ester. These predictions are borne out in practice.

The relative rates of oxidation of secondary a- and e-alcohols to ketones by chromic acid (or hypobromous acid) is the reverse of the relative rates of hydrolysis of their carboxylic esters, *i.e.,* an a-hydroxyl is more readily oxidised than an e-. The reason for this is that the rate-determining step in this oxidation is a direct attack on the hydrogen atom of the C–H bond. If the hydroxyl is axial, the hydrogen is equatorial, and vice versa; thus:

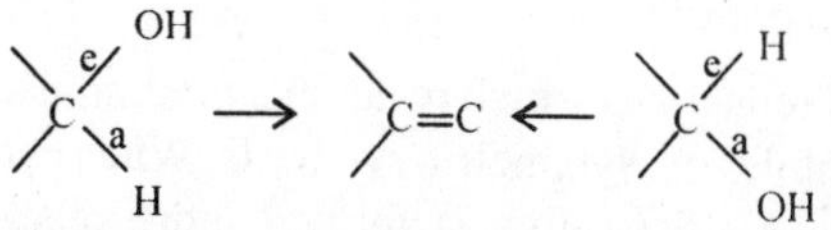

Let us now consider some chemical reactions of cyclohexanones. Enolisation would, at first sight, appear to occur equally well with either an axial or equatorial hydrogen atom on the α-carbon atom. Corey *et al.,* (1956), however, showed that it is the axial hydrogen atom which is mainly involved, and this has been attributed to the stereoelectronic factor. Reduction of the carbonyl group may occur in two ways: approach from the axial side will produce an equatorial alcohol (since the C–H bond produced is axial); and approach from the equatorial side will produce an axial alcohol (since the C–H bond formed is equatorial). Catalytic hydrogenation results predominantly in the formation of the axial alcohol. This may be explained by assuming that the ketone is adsorbed on the catalyst preferentially from the less-hindered equatorial side, and since the adsorbed hydrogen must also come from this side the product is the axial alcohol.

On the other hand, reduction with metal and acid gives predominantly the equatorial alcohol. In this case, the proton (abstracted from H_3O^+) is small and so can approach an unhindered carbonyl group equally well from either side, but since the equatorial conformation is more stable than the axial, e-alcohol is the main product. Because the predominant product depends on its relative stability, this has been referred to as an example of product development control. When the reducing reagent is aluminium isopropoxide (MPV reduction), the predominant product is now the a-alcohol. Because of the large spatial requirements in the cyclic transition state, the aluminium complex will preferentially be formed from the less hindered equatorial side. Since the predominant product depends on steric factors, this is an example of steric approach control (Dauben *et al.,* 1956).

Reduction of cyclohexanone derivatives with lithium aluminium hydride or with sodium borohydride generally produces the equatorial alcohol as the predominant product. In these reductions, the spatial requirements of the aluminium hydride and borohydride ions are much less than those of aluminium isoproxide. The product is therefore an example of product development control. If, however, the ketone is sterically hindered, *e.g,* if the ketone is 3,3-dimethylcyclohexanone, the axial alconol is the predominant product. Here we have an example of steric approach control.

All of these reductions which result in a predominant stereoisomeric product are examples of asymmetric synthesis. When cyclohexanones are halogenated, the 2-a-derivative is formed more readily than the 2-e-isomer (dipolar effect). Even so, the final product may be predominantly the 2-e-isomer due to 1,3-interactions.

Elimination reactions are also of great importance in cyclic compounds. As we have seen, in ionic E2 reactions the two groups eliminated are normally in the trans position. In cyclohexane systems this geometrical requirement is only found in trans- 1 ,2-diaxial compounds, and these compounds thus undergo ready elimination reactions. In rigid systems, *e.g,* the trans-decalin type, elimination in trans-1,2-diequatorial compounds is slower than in the corresponding diaxial compounds. cis-1,2-Compounds (in which one substituent must be axial and the other equatorial) undergo elimination reactions slowly.

The steric course of E1 reactions is more difficult to study than that of E2 reactions because of the two-stage mechanism. This makes it difficult to ascertain the geometry of the intermediates involved. The formation of the carbonium ion will be sterically accelerated if the ionising group is axial and, if a second group is eliminated to form a double bond, this second stage will also be sterically accelerated if the second group is axial (see below).

The arguments used above are satisfactory so long as we know whether the group under discussion is axial or equatorial. Since, however, the two chair forms are readily interconvertible and in equilibrium, to study these predictions experimentally it is necessary to deal with 'rigid' conformations. The t-butyl group, because of its large size, is far more stable in the e- than in the a-position (the energy difference between the two forms is about 23.4 kJ mol^{-1}; Winstein *et al.,* 1955).

Thus, almost only the e-form is present and consequently this position is 'locked'. Alternatively, the t-butyl group is referred to as an 'anchor' or 'anchoring group', and the compounds are said to be conformationally 'biased'. Hence, on the basis that the t-butyl group is equatorial, 4-

substituents must be axial when cis to the t-butyl group and equatorial when trans to this group. Working with different substituents in the 4-position with respect to the t-butyl group, various workers have confirmed the above predictions experimentally, *e.g,* it has been shown that cis-4-t-butylcyclohexanol forms esters more slowly than the trans-isomer, and similarly cis-4-t-butylcyclohexane-lcarboxylic acid is more slowly esterified and the ester more slowly hydrolysed than the trans-isomer.

Another interesting example is the case of 4-t-butylcyclohexyl tosylate. Two forms are possible, cis and trans, but because of the large bulk of the t-butyl group, this group is always equatorial. Under the same conditions (sodium ethoxide in ethanol at 70°C), the cis-form readily undergoes bimolecular-elimination (E2), but the trans- does not. Actually, the latter does undergo bimolecular elimination, but this is so slow that it is virtually undetectable in the presence of the concurrent El reaction.

Some examples of neighbouring group participation in cyclohexane systems have been described. These examples clearly show the effect of conformation on rates of reaction when anchimeric assistance is possible. Not only does conformation control the rate of reactions, but it also may affect the course of a reaction. An example is the action of nitrous acid on amines. Mills (1953) has proposed the following generalisation: When the amino-group is equatorial, the product is an alcohol with an equatorial conformation; but when the amino-group is axial, the main product is an alkene together with some equatorial alcohol. The basis of this generalisation is as follows.

The diazonium ion produced undergoes unimolecular decomposition, and if the amino group is axial, then the diazo group is axial and the carbonium ion readily undergoes elimination if there is an axial hydrogen atom on the adjacent carbon atom. At the same time a small amount of equatorial alcohol is also formed. When the amino group is equatorial, the leaving diazo group is equatorial. Since the adjacent hydrogen atoms are no longer suitably placed for attack at the back of the receding diazo group, the result is formation of equatorial alcohol only.

There are, however, exceptions to the above generalisation, *e.g,* trans-4-t-butylcyclohexylamine (e-t-Bu : e-NH2) gives mainly the trans-alcohol, but only 13 per cent of the cis-alcohol (e-t-Bu : a-OH) and 10 per cent of alkene. Also, cis-4-t-butylcyclohexylamine (e-t-Bu: a-NH,) gives a mixture of the corresponding alcohols and predominantly the alkene (Huckel *et al.,* 1963). The explanation for these results is uncertain.

As we have seen, although there is a preferred form in cyclohexane derivatives, the energy barrier of interconversion between the preferred and less stable form is too low to permit their being distinguished by the classical methods of stereochemistry. This predominance of the preferred form holds good at room temperature (or below). At higher temperatures, or during the course of a chemical reaction, the preponderance of the preferred Corm may be reduced. In chemical reactions, it may be possible for the reaction to proceed more readily through the less stable conformation because it is this one which more closely approaches the geometry of the transition state. An example of this type is chlorocyclohexane. As we have seen, the preferred form is the equatorial conformation. This compound, on treatment with ethanolic potassium hydroxide, undergoes dehydrohalogenation to form cyclohexene. Since trans elimination is preferred, and since the rate of interconversion of the conformers is much faster than the rate of the elimination, elimination can be expected to occur via the axial conformation.

Systems of this type, *i.e.*, conformers of cyclohexane derivatives in equilibrium, have been studied from the point of view of determining the conformational equilibrium constant and the reactivities of both conformations. One example is the acetylation of cyclohexanol with acetic anhydride in pyridine at 25°C (Eliel *et al.*, 1957).

If k is the overall observed rate of acetylation and ke and ka are the rates of acetylation of the equatorial and axial conformers, respectively, then (Winstein *et al.*, 1955):

$$k = k_eN_e + k_aN_a$$

where N_e and N_a are the mole fractions in the equatorial and axial conformations, respectively. Also

$$N_e + N_a = 1 \text{ and } K = N_e/N_a$$

where K is the conformational equilibrium constant. The value of k can be readily obtained but the difficulty is to obtain the values of k_e and k_a because the two conformations are too readily interconvertible to be studied separately. However, the required values of k_e and k_a have been determined by using cis- and trans-4-t-butylcyclohexane derivatives, the assumption being made that the t-butyl group is exclusively equatorial. This is essentially true provided the other group is not comparable in size with the t-butyl group; this is usually the case. In this way, the conformations of the cis- and trans-isomers are known and are therefore fixed.

According to Eliel, k for cyclohexanol is 8.37×10^{-5}, k_a for cis-4-t-butylcyclohexanol (e-t-Bu, a–OH; Na = 1, $N_e = 0$) is 2.89×10^{-5}, and k_e for trans-4-t-butylcyclohexanol (e-t-Bu, e-OH; $N_a = 0$, $N_e = 1$) is 10.65×10^{-5} (1 mol^{-1} s^{-1}). Hence:

$$8.37 = 10.65N_e + 2.89N_a$$
$$= 10.65N_e + 289(1 - N_e)$$
$$= N_e(10.65 - 2.89) + 2.89$$
$$\therefore \quad 7.76N_e = 5.48$$
$$\therefore \quad N_e = 5.48/7.76 = 0.71$$
$$\therefore \quad N_a = 0.29$$

Hence $K = 0.71/0.29 = 2.45$

Also, since $\Delta G° = -RT \text{ In } K$

$$\therefore \quad \Delta G° = -2.2 \text{ kJ mol}^{-1}$$

i.e., the free-energy difference between equatorial and axial hydroxyl is -2.2 kJ mol^{-1} (the equatorial form is more stable than the axial).

Eliel (1960), independently of Winstein, derived the equivalent relationship:

$$k = (k_eK + k_a)/(K + 1)$$

This equation, when solved for K, gives:

$$K = (k_a - k)/(k - k_e)$$

The equivalence of the Winstein and Eliel relationships is readily established as follows:

$$K = N_e/N_a$$
$$K + 1 = (N_e + N_a)/N_a = 1/N_a$$

Substituting these values in Eliel's equation, we get

$$k = (k_eN_e/N_a + k_a)/(1/N_a)$$
$$= k_eN_e + k_aN_a$$

Conformational equilibrium constants (and hence free-energy differences) may be calculated more directly from equilibrium reactions. These reactions can occur for various types of cyclohexane derivatives, the most important group being that involving the interconversion of isomers via an enol or enolate ion. Thus, for example, either cis- or trans-

4-t-butyl-2-methylcyclohexanone is converted into an equilibrium mixture of the two isomers in which the cis-isomer predominates. This is to be expected on the basis that the e-methyl conformation is more stable than the a-form.

Not only do the values of free-energy differences show which is the more stable conformer, they also give some indication of the 'size' of the groups, since it can be anticipated that the larger the size, the higher will be the free-energy difference. These values, however, depend on temperature and the nature of the solvent. Some values are (kJ mol^{-1}): halogen, 0–2; OH, OR, 2–4; CO_2H, CO_2R, 4–6.2; Me, Et, NH_2, 6.2–8.3; Pr, isoPr, NMe_2, 8.3–10.4; Ph, 13; t-Bu, > 16.7.

Doubts have been expressed about the validity of conformational energies computed by means of the kinetic methods (inter alia, Kwart *et al.*, 1964, 1967).

Conformational equilibrium constants may also be determined by physical methods: i.r., u.v., NMR, ORD, etc. Thus, *e.g,* in the infrared method it is necessary to find an absorption frequency characteristic of the equatorial and the axial conformation. This may be done by using cis- and trans-t-butyl derivatives containing the group under investigation (see above). Then, on the assumption that the molar extinction coefficients are the same in the compound being investigated and its corresponding t-butyl derivative, the ratio of the intensities of the bands in these two compounds gives the mole fraction of the equatorial or axial conformation. In the NM R method, if δ is the chemical shift of the proton of the CHOH group in cyclohexanol and δ_e (axial proton) and δ_a (equatorial proton), those of the corresponding trans- and cis-4-t-butyl derivatives (e-OH and a-OH, respectively), then, on the assumption that the 4-t-butyl group has no effect on the chemical shift:

$$\delta = N_e\delta_e + N_a\delta_a$$

From this the ratio $N_e/N_a = K$ can be calculated.

An alternative approach uses the fact that the values of coupling constants vary with the dihedral angle between the coupled protons.

Rearrangements Involving Cyclohexanes: Most cyclohexane derivatives can undergo 1,2-shif.s, but the course of the reaction depends on a number of factors. If the leaving group Z is axial and there is a group Y attached to an adjacent carbon atom which is also axial, *i.e.*, trans, and capable of migration, *e.g,* methyl, hydrogen, then the spatial requirements are satisfied and so a 1,2-shift can occur with the formation of a cyclohexane derivative (see III). If the leaving group Z is equatoria and Y axial (*i.e.*, cis), then these two groups no longer satisfy the spatial requirements.

However, an adjacent carbon atom of the ring system does satisfy the spatial requirements of the 1,2-shift, and this occurs with the formation of a cyclopentane derivative (see IV). These rearrangements are also examples of neighbouring group participation, but this may be absent in certain cases. These principles may be illustrated by the deamination of 2-aminocyclohexanol (McCasland, 1951).

The trans-isomer, on treatment with nitrous acid, forms cyclopentanecarboxaldehyde. Both groups are equatorial and since this conformer is much more favoured than the diaxial conformer, the molecule reacts in the former conformation. Thus: On the other hand, the cis-isomer gives cyclopentanecarboxaldehyde and cyclohexanone under the same conditions. In this isomer, the two e,a-conformations contain about the same amount of energy and hence are in equilibrium of approximately equal populations.

CYCLOALKENES

A very important problem in connection with elimination reactions is the conformation of cyclohexene. The presence of the double bond causes the ring to assume the half-chair conformation (atoms 1, 2, 3 and 6 are in one plane). This has been demonstrated by means of X-ray and electrondiffraction studies of some cyclohexene derivatives, *e.g*, 3,4,5,6-tetrachlorocyclohexene. The hydrogen atoms at positions 4 and 5 are the normal axial and equatorial types, but those at 3 and 6 are referred to as quasi-axial (pseudo-axial) (a′) and quasi-equatorial (pseudo-equatorial) (e′); the latter is the more stable.

Since the addition of halogen and halogen acids (by the polar mechanism) normally produces the trans-product, it can be anticipated that the predominant product will be the diaxial one in a rigid system. In a mobile system, however, if the diaxial is less stable than the diequatorial form, then the equilibrium will shift to the latter, *e.g*, when bromine adds to cyclohexene, the product is trans-l,2dibromocyclohexane, and the diaxial and diequatorial forms are present in about equal amounts (Hassel *et al.*, 1947).

The two C–Br bonds produce strong dipole-dipole interactions and so partly offset the 1,3-interactions. An interesting point in this connection is that Berti *et al.*, (1963) have shown that the addition of bromine to cyclohexene in the presence of cinchonine or cinchonidine gives optically active trans-1,2-dibromocyclohexane. Epoxidation occurs by a cis-mechanism, and it has been shown by electron diffraction measurements that the half-chair conformation in cyclohexene is retained in 1,2-epoxycyclohexane (Ottar, 1947).

A particularly interesting case of the conformation of a cycloalkene is cyclo-octene. The transisomer is not superimposable on its mirror image, and has been resolved by Cope *et al.*, (1963). It is the first cyclic alkene that has been found to contain molecular asymmetry.

BRIDGED-RING SYSTEMS

A number of these systems occur in various bicyclic terpenes. On the other hand, many complicated bridged-ring systems have been synthesised recently. Tamelen *et al.*, (1963) have prepared bicyclo[2,2,0] hexadiene (I); this is the 'Dewar benzene'. It has a half-life in pyridine of two days at room temperature, and when heated to 90°C it is converted into benzene.

Adamantane (II) was first isolated from petroleum (Landa *et al.*, 1933), but it has now been synthesised in various ways, *e.g*, dicyclopentadiene is hydrogenated and the product isomerised by aluminium chloride (Schleyer, 1957). It is composed of three fused cyclohexane chair conformations; its structure is rigid but strainless. Twistane (III), the twist-boat isomer of adamantane, has been synthesised by Whitlock, Jr. (1962). Other compounds which have been prepared are, *e.g*, barrelene (IV) [Zimmerman *et al.*, 1960] and cubane (V) [Eaton *et al.*, 1964].

Catenanes

These are large ring compounds consisting of two interlocked rings. Wasserman (1960) prepared (I), and this compound, in the form of a hydrocarbon, may be represented as (II).

Each ring can be described independently of the other, and since they share no bonds, they are chemically independent. To convert the catenane into the corresponding non-interlocked pair of rings, it is necessary to break a bond in one of the rings. This bond is not a property of a particular pair of atoms in a catenane hydrocarbon, since all atoms are equivalent in each ring.

Thus, the bond to be broken is a property of the two complete rings. This bond is called a topological bond, and the relationship of the catenane to its pair of 'unlocked' rings is called topological isomerism. DNAs have been shown to exist as catenanes.

(III), which is a trefoil, is an example of a single chain that is knotted. It is a topological isomer of the corresponding unknotted cyclohexane, and the two isomers can be interconverted only by breaking a C—C bond.

The trefoil (IV) has been synthesised by Schill (1969).

Rotaxanes: These are circular macromolecules which are threaded by a spindle whose ends consist of groups of such bulk that the 'ring' cannot 'fall off' the spindle (see V). Various rotaxanes have now been synthesised, *e.g,* (VI) (Harrison, 1972).

A very interesting example of chemical topology is 3-methyl-5-bromoadamantanecarboxvlic acid (VII).

This may be regarded as the formal analogue of 2-bromopropionic acid (VIII) in which the centre of chirality (denoted by the dot) is at the 'unoccupied' centre of (VII). This compound has been resolved by McKervey *et al.,* (1969).

SMALL RINGS

We have already discussed the six-membered ring, and the principles described may be applied to the other rings. Cyclopropane, since it is a three-point system, must be planar. Large rings are puckered, contain very little strain, and the chemistry of their derivatives closely resembles that of the aliphatic analogues.

Cyclobutane is a four-point system and so could be either planar or puckered. If it were planar then, apart from angle strain, there could be strain due to eclipsing of adjacent hydrogen atoms and to bond opposition (I). On the other hand if the ring were puckered, then eclipsing and bond opposition strains would be diminished, but angle strain would be increased. As we saw in bond angle deformation is brought about comparatively easily, and so the puckered form might therefore be more stable than the planar. This has actually been shown to be the case; electron diffraction (Bastiansen *et al.,* 1961) and spectroscopic and thermodynamic measurements (Pitzer *e. al.,* 1953) have shown that cyclobutane is puckered. Allinger *et al.,* (1962) showed, by thermodynamic calculations, that methy! cis-3-methylcyclobutane-carboxylate (II) is more stable than the trans-isomer by 1.3 kJ mol^{-1}. This is in agreement with a puckered structure, but Allinger *et al.*, (1965) found that dimethyl trans-cyclobutane-l,3-dicarboxylate (III) is more stable than the cisisomer by 0.42 kJ mol^{-1}. The greater stability of the trans-isomer appears to be due to dipole-dipole interactions. On the other hand, the trans-isomer of cyclobutane-l,3-dicarboxylic acid has been shown, by X-ray analysis, to have a planar ring in the solid state, whereas the cis-isomer has a puckered ring (Margulis *et al.*, 1967). However, it should be noted that X-ray analysis on single crystals may give a shape of the molecule dominated by packing effects within the crystal and not the shape of the molecule in different circumstances, *e.g,* in solution.

COMMON RINGS

Cyclopentane could be planar or puckered, and the arguments used for cyclobutane suggest that the puckered form might be the more stable one. However, since cyclopentane is a five-point system, it can have four carbon atoms in one plane and the remaining one outside (envelope or CS form) or three in one plane and the other two outside (half-chair or C_2 form).

Electron diffraction measurements of cyclopentane indicate puckering (Bastiansen *et al.*, 1961). The puckering is not fixed, but 'rotates' round the ring, *i.e.*, each carbon atom oscillates in a direction perpendicular to the average plane of the ring, *i.e.*, no particular carbon atom is always out of plane, but each one takes up this conformation. This effect is called pseudorotation. Pseudorotation occurs in cyclopentane because the potential energy barriers are very low, but when the ring carries one or more substituents, the energy barriers may now be high enough to inhibit pseudorotation. The projected angle for cis-1 ,2,-substituents in cyclohexane is 60°, whereas that for cis-1,2 (and 1,5) is about 46° and that for cis-2,3 (and 4,5) is about 29° in the envelope form of cyclopentane (Pitzer *et al.*, 1959).

Also, since cis-3,4. substituents are eclipsed, instability due to this eclipsing can be relieved by the compound assuming the half-chair conformation. For cyclopentane itself the two conformations differ very little in energy content (Pitzer *et al.*, 1959). The conformation of the ring, however, in cyclopentane derivatives depends on the nature of the substituents. Calculations by Pitzer *et al.*, (1949) have shown that cis-1,3-dimethylcyclopentane is more stable by 2.1 kJ mol^{-1} than the trans-isomer; this is in agreement with the envelope conformation. On the other hand, cyclopentanone is more stable in the half-chair form; the remaining hydrogens possess a minimum of steric repulsions (due to maximum staggering in this form).

Pitzer *et al.,* (1959) have also carried out calculations on heterocyclic 5-membered rings, *e.g,* pyrrolidine, tetrahydrofuran, etc., and conclude that these compounds are also in the half-chair form (the hetero-atom replaces the CO group in cyclopentanone).

Cycloheptane has, according to the calculations of Hendrickson (1961), two stable conformations, the twist-chair and the slightly less favoured twist-boat conformation, both of which are flexible, and interconversion of the equatorial and axial bonds in the twist-chair can be achieved by pseudorotation. Cycloheptatriene, a compound obtained from tropine, exists in a boat-shaped conformation (Traetterberg, 1964).

Medium Rings: Several conformations have been proposed for cyclo-octane: the extended crown (IV) (Mez, 1960) and the 'saddle' conformation (V) (Dale *et al.,* 1964). Strain energy minimisation calculations by Hendrickson (1964) and Wiberg (1965) suggest that neither form is clearly favoured energetically. Dunitz *et al.,* (1966) have examined, by X-ray analysis, cyclo-octane-1,2-transdicarboxylic acid, and showed that the conformation of this molecule is (VI), the 'boat-chair' form (so named by Hendrickson).

Cyclo-octatetraene (VII) exists in the tub form (Treibs, 1950). This compound is of particular interest because bond fluctuation, called fluxionalism, has been observed and measured by NMR spectroscopy in, *e.g,* fluorocyclo-octatetraene. Its NMR spectrum shows a broad band at room temperature because, at this temperature, bond fluctuation and ring inversion occur rapidly. At-65°C, however, these processes are sufficiently slowed down for the spectrum to show a doublet due to coupling of $H_{(2)}$ with F in (VIIa) or $H_{(8)}$ with F in (VIIb). Cyclononane has the flexible symmetrical form (VIII) (Hendrickson, 1964).

Cyclodecane: X-ray analysis studies on some cyclodecane derivatives have shown that the conformation is (IX) (Dunitz *et al.,* 1960). This is derived from two chair conformations of cyclohexane joined by 1,3-axial bonds. It is these more complicated conformations which make the medium rings so different from large rings in many chemical and physical properties. The anomalous behaviour has been explained in terms of bond opposition strain, angle strain, and steric strain, in the last of which a very important contributing factor is the inter actions between atoms on opposite sides of the ring (see (IX)). This type of interaction has been designated transannular interaction, and this produces transannular strain.

Examples of anomalous reactions are, *e.g,* the formolysis of epoxycyclodecane (from transcyclodecene) to give trans-cyclodecane-1,6-diol (Prelog *et al.,* 1952). The structures have been written in the conventional manner; the reaction occurs by a transannular hydride shift: Addition of bromine to cis- and trans-cyclodecene gives, respectively, cis- and trans-l,6-dibromocyclodecane (Sicher *et al.,* 1961). Another example is the catalytic reduction of cyclodecane-i,6dione (Kosover *et al.,* 1961).

Large Rings: It is also of interest to discuss strain in alicyclic compounds in the following way. The heat of combustion per CH_2 group in acyclic n-hydrocarbons is 659.0 kJ. The values for alicyclic hydrocarbons depend on the size of the ring: 3, 696.6; 4, 6853; 5, 664.0; 6, 659.0; 7, 661.9; 8–11, 661.1–665.3; 12–, 656.0–661.1 kJ mol^{-1}. Thus, strain is a minimum in cyclohexane, and strain relative to cyclohexane may therefore be estimated, *e.g,* the difference per CH_2 for 5- and 6-rings is 5.0 kJ, and so the strain in cyclopentane is $5 \times 5.0 = 25$ kJ mol^{-1}.

The strains in medium rings may also be calculated in this way. The values are about 25.1 kJ mol^{-1} for cycloheptane, about 37.7–41.8 for cyclo-octane, and about 50.2–54.4 for cyclononane to cycloundecane. From cyclododecane onwards, the value of CH_2 groups is close to that of the acyclic hydrocarbons, *i.e.,* these large rings are strainless and have been shown (by X-ray analysis) to consist of two parallel chains which are puckered.

I-Strain: In general, many reactions of medium rings are either faster or slower than those of the other ring systems, *e.g,* the rate of hydrolysis of medium-ring chlorides and p-toluenesulphonates is faster, reduction of medium-ring ketones by sodium borohydride slower, dissociation of these ketone cyanohydrins greater, etc. Many of these differences in reactivity can be explained in terms of steric strain. The total ring strain in cyclic systems has been called I-strain (internal strain; Brown, 1951, 1956).

The contributing factors to strain in alicyclic hydrocarbons, as we have seen, are steric repulsion, bond opposition forces, and angle deformation. In medium rings, however, there are also transannular interactions. The difficulty is the estimation of these contributions. It is assumed, however, that in small rings, the major source of strain is angle distortion; in 5- and 7-rings, bond opposition forces; and in medium rings all four factors play a part, with transannular interactions particularly prominent.

Changes in I-strain may be considered in terms of changes from sp^2 to sp^3 hybridisation, or vice versa, at the site of reaction. Brown assumed that if the change $sp^2 \rightarrow sp^3$ is accompanied by a decrease in I-strain, reactions involving this change are facilitated. Also, reactions involving the change $sp^3 \rightarrow sp^2$ are hindered. On the other hand, in those situations where the change $sp^2 \rightarrow sp^3$ is accompanied by an increase in I-strain, reactions are hindered, and those for $sp^3 \rightarrow sp^2$ are facilitated.

Whether a change in hybridisation facilitates or hinders a reaction depends on the size of the ring. For a simple illustration, let us consider the hydrolysis of 1-chloro-l-methylcyclopropane by the S_N1 mechanism. When the carbonium ion is produced by ionisation ($sp^3 \rightarrow sp^2$), the bond angle strain [(109.5° – 60°)/2] would change to [(120° – 60°)/2].

Hence, internal strain is increased and therefore the rate of hydrolysis can be expected to be slower than for any other similar ring compound (in which angle strain is less because the bond angle is greater than 60°). This is so in practice.

2

Nomenclature of Geometrical Isomers

INTRODUCTION

When geometrical isomerism is due to the presence of one double bond in a molecule, it is easy to name the geometrical isomers if two groups are identical, *e.g,* in molecules (I) and (II), (I) is the cis-isomer and (II) the trans; similarly (III) is cis and (IV) is trans. When, however, all four groups are different, nomenclature is more difficult. In this case it has been suggested that the prefixes cis and trans should indicate the disposition of the first two groups named, *e.g,* the two stereoisomers of 1-bromo-1-chloro-2-iodoethylene, (V) and (VI); (V) is cis-l-bromo-2-iodo-l-chloroethylene or trans-1-chloro-2-iodo-l-bromoethylene: (VI) is cis-l-chloro-2-iodo-l-bromoethylene or trans-lbromo-2-iodo-l-chloroethylene. On the other hand, since this method of nomenclature

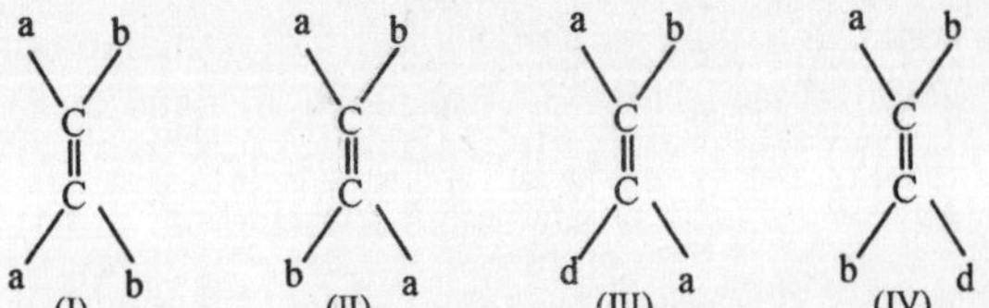

(I) cis　(II) trans　(III) cis　(IV) trans

usually deviates from the rule of naming groups in alphabetical order, it has been suggested that the groups corresponding to the prefix cis or trans should be italicised, thus (V) may be named cis-l-bromo-lchloro-2-iodoethylene and (VI) trans- I-bromo-I-chi oro-2-iodoethylene. This method, it must be admitted, would offer difficulties when the names are spoken.

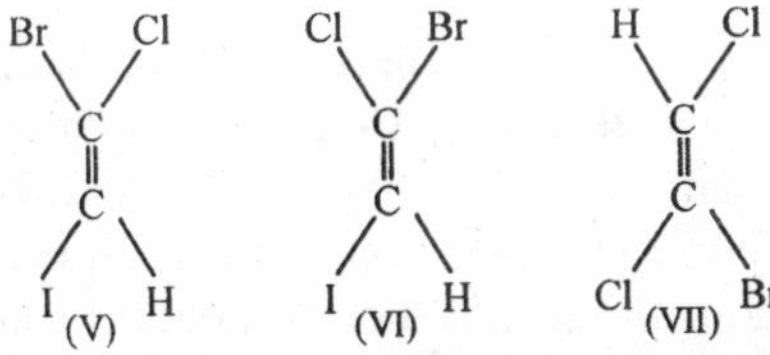

(V) (VI) (VII)

These difficulties have now been overcome by the introduction of a new system of nomenclature. Let us consider the molecule UXC=CYZ. The groups U and X, and the groups Y and Z are now arranged in order of precedence in accordance with the Sequence Rule in the R-S system.

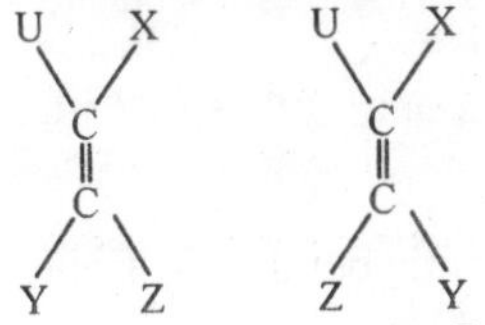

seqcis or Z seqtrans or E

Sequence:

U > X

Y > Z

Let us suppose that the order of precedence is U > X and Y > Z. Then, if the groups U and Y (both of higher precedence) are on the same side, the configuration of the alkene is seqcis, and if they are on opposite sides, seqtrans. Thus, *e.g,* since the order of precedence of Br, Cl, I, and H is Br > Cl and I > H, (V) is seqcis and (VI) is seqtrans.

An alternative scheme-the E–Z system–uses the symbols Z (German: zusammen = together) and E (German: entgegen = opposite). Thus, (V) is (Z)-l-bromo-l-chloro-2-iodoethylene, and (VI) is the corresponding (E)-isomer. These symbols correspond to seqcis (Z) and seqtrans (E), but do not necessarily correspond to cis and trans (in the earlier nomenclature), *e.g,* (VII) is trans1,2-dichloro-l-bromoethylene, but by the E–Z system is the (Z)-isomer (H < Cl, Cl < Br; hence seqcis or Z).

The naming of faces of compounds containing double bonds has been discussed.

Some pairs of geometrical isomers have trivial names, *e.g,* maleic and fumaric acids, angelic and tiglic acids, etc. Sometimes the prefix iso has been used to designate the less stable isomer, *e.g,* crotonic acid (trans-isomer) and isocrotonic acid. The use of iso in this connection is undesirable since it already has a specific meaning in the nomenclature of alkanes. The prefix allo has also been used to designate the less stable isomer (cis), *e.g,* allocinnamic acid.

When geometrical isomers contain two or more double bonds, nomenclature may be difficult, *e.g,* (VIII). In this case the compound is considered as a derivative of the longest chain which contains the maximum number of double bonds, the prefixes cis and trans being placed before the numbers indicating the positions of the double bonds to describe the relative positions of the carbon atoms in the main chain; thus (VIII) is 3-isopropylhexa-cis-2,cis-4-diene.

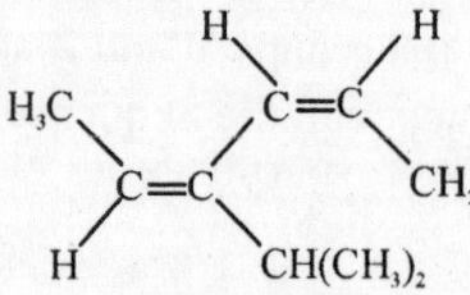

(VIII)

If a compound has two double bonds, *e.g,* CHa=CH–CH=CHb, four geometrical isomers are possible:

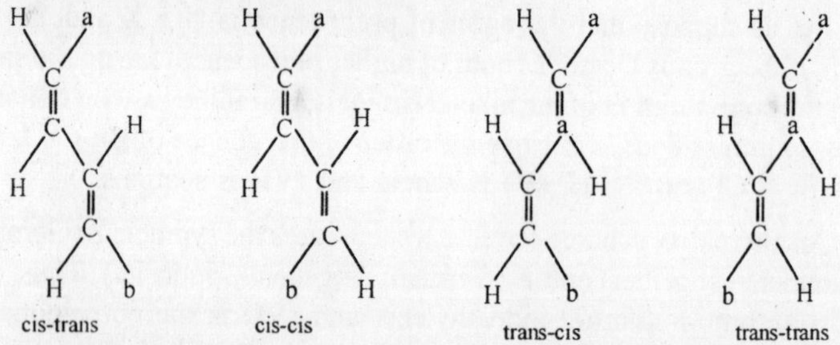

The number of geometrical isomers is 2^n, where n is the number of double bonds; this formula applies only to molecules in which the ends are different. If the ends are identical, *e.g,* CHa=CHCH=CHa, then the number of stereoisomers is $2^{n-1} + 2^{p-1}$, where $p = n/2$ when n is even, and $p = (n + 1)/2$ when n is odd (Kuhn *et al.,* 1928).

Geometrical isomerism is also exhibited by cumulenes provided that the number of adjacent double bonds is odd, *e.g,* 1,4-di-3-nitrophenyl-1,4-diphenylbutatriene exists in two forms (Kuhn *et al.,* 1959). On the other hand, cumulenes containing an even number of double bonds exhibit optical isomerism.

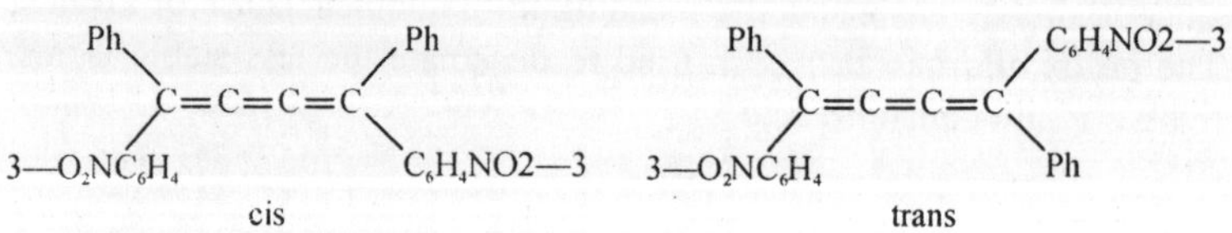

It has previously been pointed out that geometrical isomers, as such, are not optically active, but if they contain chiral centres, then they will also exhibit optical isomerism. In such circumstances, the presence of the double bond leads to a larger number of optical isomers. This type of isomerism is known as geometrical enantiomerism. If Z^+ and Z^- represent mirror-image forms of an asymmetric group, then a molecule of the type CZ–CZ will exist as one pair of enantiomers and one meso-form (CZ^+–CZ^+, CZ^-–CZ^-, and CZ^+–CZ^-). If we now consider a molecule containing these two Z groups attached to an olefinic carbon atom as shown, then four optically active forms are now possible.

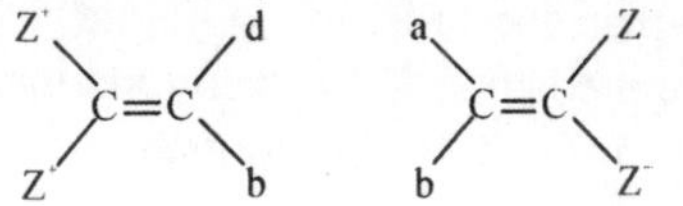

pair of enantiomers

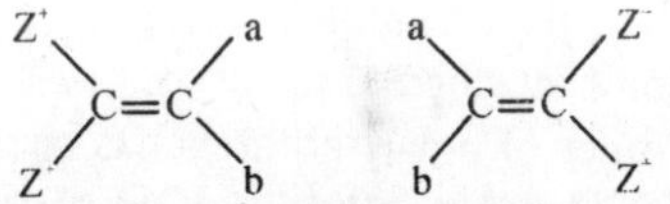

pair of enantiomers

It might also be noted that geometrical isomers are now classified as diastereoisomers. According to the new definition, diastereoisomers are any stereoisomers which are not enantiomers of each other, *i.e.,* the restriction that diastereoisomers are optical isomers has been dropped. Thus this type of isomerism–cis-trans or geometrical–is a sub-class of the general phenomenon of diastereoisomerism.

CORRELATIONS OF SIGN AND MAGNITUDE OF ROTATION WITH ABSOLUTE CONFIGURATION

Brewster (1959, 1961) has devised an empirical correlation (for rotation with the sodium D line) and has used a number of general rules for this purpose. These general rules are based on the following hypothesis: A centre of optical activity can usefully be described as an asymmetric screw pattern of polarisability. This screw pattern, however, may arise in one of two ways or as a combination of both of them:

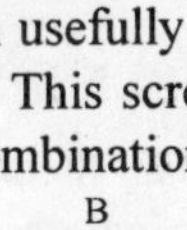

(I)

(i) **Atomic asymmetry :** If the tetrahedral system XABCD has the absolute configuration shown in (I), it is dextrorotatory when the order of polarisability of the groups is A > B > C > D. Provided the groups A, B, C, and D are atoms or simple groups (which do not introduce conformational problems), the rules for the prediction of the sign (and magnitude) of rotation are readily applied.

(ii) **Conformational asymmetry :** In this case, the polarisability is caused by the conformation of the groups in the molecule. When this is present, its contribution to the molecular rotation is usually larger than that due to atomic asymmetry. Application of the rules for the prediction of the sign (and magnitude) of rotation is far more difficult for conformational asymmetry than for atomic asymmetry.

Atoms and groups can be arranged in order of decreasing polarisability, the individual polarisabilities being derived from the atomic refractions of the atoms attached to the asymmetric carbon atom. In this way, the following order of polarisabilities has been obtained: I > Br > SH > Cl > Ph = CO2H > Me > NH2 > OH >H>D>F.

Let us now use a-phenylethyl chloride (II) to illustrate the application of the rules for atomic asymmetry. The absolute configuration of (R)-α-phenylethyl chloride has been shown to be (II). Reference to the order

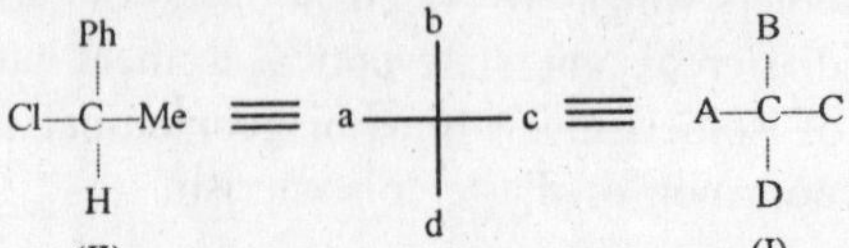

(II) (I)

of the polarisabilities of the groups gives the configuration (I). Therefore this enantiomer is predicted to be dextrorotatory. This is the case in practice (the (R)-form is the dextrorotatory enantiomer). If we referred to a table of values of polarisabilities, we would also find that the magnitudes of the predicted and observed rotations are in fair agreement.

ROTATION ABOUT A DOUBLE BOND

We have already seen that, theoretically, there is always some opposition to rotation about a single bond and that, in many cases, the opposition may be great enough to cause the molecule to assume some preferred conformation. When we consider the problem of rotation about a double bond, we find that there is always considerable opposition to the rotation. Let us first consider the simple case of ethylene; the energy changes in the molecule when one methylene group is rotated about the carbon-carbon double bond with the other methylene group at rest. Thus there are two identical favoured positions (one at 0° and the other at 180°), and the potential energy barrier is 1674 kJ mol^{-1}. The examination of many olefinic compounds has shown that the potential energy barrier for the C=C bond varies with the nature of the groups attached to each carbon, *e.g,*

$$CH_2=CH_2,\ 167.4\ kJ\ mol^{-1}$$

$$C_6H_5CH=CHC_6H_5,\ 179.1\ kJ\ mol^{-1}$$

$$CH_3CH=CHCH_3,\ 75.3\ kJ\ mol^{-1}$$

$$HO_2CCH=CHCO_2H\ 66.1\ kJ\ mol^{-1}$$

Let us consider the case of maleic and fumaric acids in more detail. It can be seen from the favoured positions, with the trans-form more stable than the cis, the energy difference between the two being 25-29 kJ mol^{-1}. The conversion of the trans to the cis requires 66.1 kJ energy, but the reverse change requires about 42 kJ.

THE NATURE OF DOUBLE BONDS

In the foregoing account of geometrical isomerism, the distribution of the carbon valencies was assumed to be tetrahedral (as postulated by van't Hoff). According to one modem theory, the four valency bonds of a carbon atom are distributed tetrahedrally only in saturated compounds. In such compounds the carbon is in a state of tetrahedral hybridisation, the four sp^3 bonds being referred to as α-bonds. In olefinic compounds, however, the two carbon atoms exhibit the trigonal mode of hybridisation. In this condition there are three coplanar valencies (three σ-bonds produced from sp^2 hybridisation), and the fourth bond (π-bond) at right angles to the trigonal hybrids. π-Bonds, which are weaker than σ-bonds, tend to overlap as much as possible in order to make the bond as strong as possible.

Maximum overlap is achieved when the molecule is planar, since in this configuration the two p_z orbitals are parallel. Distortion of the molecule from the planar configuration decreases the overlap of the π-electrons, thereby weakening the π-bond; and this distortion can only be effected by supplying energy to the molecule. It is therefore this tendency to produce maximum overlap of the π-electrons in the π-bond that gives rise to resistance of rotation about a 'double' bond. For simplicity we shall still represent a 'double' bond by the conventional method, *e.g,* C=C, but it should always be borne in mind that one of these bonds is a σ-bond (sp^2 bond), and the other is a π-bond perpendicular to the a-bond. It is these π-electrons (mobile electrons) which undergo the electromeric and resonance effects. They are held less firmly than the π-electrons and are more exposed to external influences; it is these π-electrons which are responsible for the high reactivity of unsaturated compounds. In compounds containing a triple bond, *e.g,* acetylene, the two carbon atoms are in a state of digonal hybridisation; there are two σ-bonds (sp bonds) and two π-bonds (one p_y and one p_z orbital), both perpendicular to the σ-bonds which are collinear.

The above treatment of the double (and triple) bond is in terms of sp^2 (and sp) hybridisation and π-bonds. It is still possible, however, to use sp^3 hybridisation to describe carbon-carbon multiple bonds; this treatment gives rise to 'banana-shaped' orbitals, *i.e.,* 'bent' bonds (Fig. 1.4):

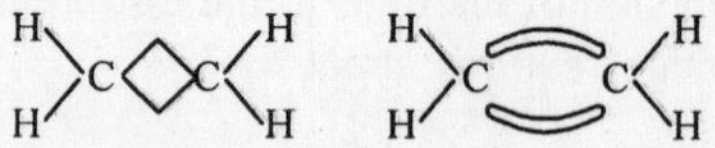

This method of approach still produces a 'rigid' molecule, and so again there is no free rotation about the double bond.

Quantum mechanical arguments show that both methods of representing these bonds are equal to each other, each method having certain advantages. The σ-π bond method is more convenient for describing transitions from one state into another, whereas the bent bond method is more convenient for describing electron distribution in a molecule.

THE RACEMIC MODIFICATION

The racemic modification is an equimolecular mixture of a pair of enantiomers, and it may be prepared in several ways.

(i) Mixing of equimolecular proportions of enantiomers produces the racemic modification.

(ii) Synthesis of chiral compounds from achiral compounds always results in the formation of the racemic modification. This statement is true only if the reaction is carried out in the absence of other optically active compounds or circularly polarised light.

(iii) **Racemisation :** The process of converting an optically active compound into the racemic modification is known as racemisation. The (+)- and (–)-forms of most compounds are capable of racemisation under the influence of heat, light or chemical reagents. Which agent is used depends on the nature of the compound, and at the same time the ease of racemisation also depends on the nature of the compound, *e.g,*

(a) Some compounds racemise so easily that they cannot be isolated in the optically active forms.

(b) A number of compounds racemise spontaneously when isolated in optically active forms.

(c) The majority of compounds racemise with various degrees of ease under the influence of different reagents.

(d) A relatively small number of compounds cannot be racemised at all.

When a molecule contains two or more asymmetric carbon atoms and the configuration of only one of these is inverted by some reaction, the process is then called epimerisation.

Many theories have been proposed to explain racemisation, but owing to the diverse nature of the structures of the various optically active compounds, one cannot expect to find one theory which would explain the racemisation of all types of optically active compounds. Thus we find that a number of mechanisms have been suggested, each one explaining the racemisation of a particular type of compound.

A number of compounds which are easily racemisable are those in which the asymmetric carbon atom is joined to a hydrogen atom and can undergo tautomeric change. Let us consider the case of keto-enol tautomerism: In the keto-form (I) the carbon joined to the hydrogen atom and the oxo group is asymmetric; in the enol-form, (II), this carbon atom has lost its asymmetry. When the enol-form reverts to the keto-form, it can do so to produce the original keto molecule (I), but owing to its symmetry, the enol-form can produce equally well the keto-form (III) in which the configuration of the asymmetric carbon atom is opposite

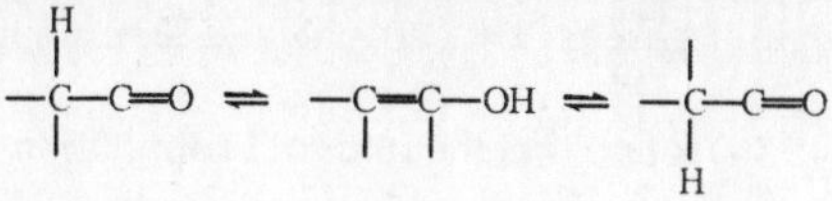

(I) (II) (III)

to that in (I). Thus racemisation, according to this scheme, occurs via the enol-form, *e.g,* (-)-lactic acid is racemised in aqueous sodium hydroxide, and this change may be formulated:

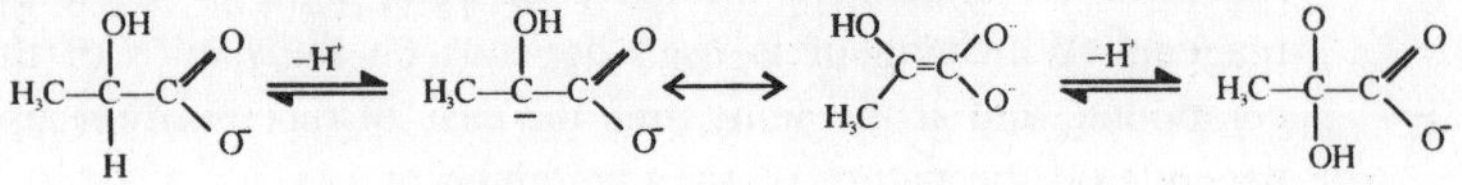

(-)

There is a great deal of evidence to support this tautomeric mechanism. When the hydrogen atom joined to the asymmetric carbon atom is replaced by some group that prevents tautomerism (enolisation) then racemisation is also prevented (at least under the same conditions as the original compound), *e.g,* mandelic acid, $C_6H_5CHOHCO_2H$, is readily racemised by warming with aqueous sodium hydroxide. On the other hand, atrolactic acid, $C_6H_5C(CH_3)(OH)CO_2H$, is not racemised under the same conditions; in this case keto-enol tautomerism is no longer possible (*i.e.,* formation of the intermediate carbanion is not possible).

Racemisation of compounds capable of exhibiting keto-enol tautomerism is catalysed by acids and bases. Since keto-enol tautomerism is also catalysed by acids and bases, then if racemisation proceeds via enolisation, the rates of racemisation and enolisation should be the same.

This relationship has been established by means of kinetic studies, *e.g,* Bartlett *et al.,* (1935) found that the rate of acid-catalysed iodination of 2-butyl phenyl ketone was the same as that of racemisation in acid solution. This is in keeping with both reactions involving the rate-controlling formation of the enol.

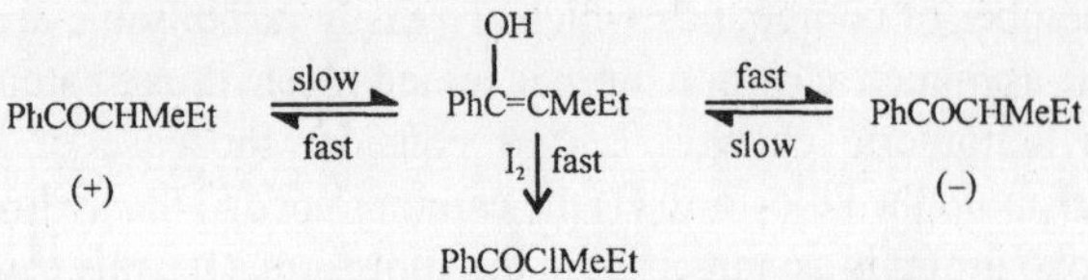

On the other hand, on the basis that the rate-determining step in base-catalysed enolisation and racemisation is the formation of the enolate ion, then the two processes will also occur at the same rate.

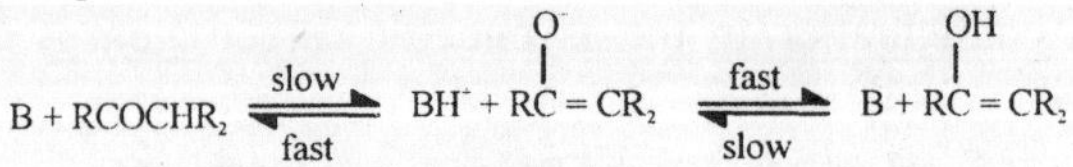

Hsu *et al.,* (1936) found that the rates of bromination and racemisation (in the presence of acetate ions) of 2-o-carboxybenzyl-l-indanone were identical.

Further support for this mechanism is the work of Ingold *et al.*, (1938) who showed that the rate of racemisalon of (+)-2-butyl phenyl ketone in dioxan-deuterium oxide solution in the presence of NaOD is the same as the rate of deuterium exchange. This is in keeping with the formation of the enolate ion (or carbanion), which is common to both reactions.

There are many compounds containing an asymmetric carbon atom which can be racemised under suitable conditions although there is no possibility of tautomerism. A number of different types of compounds fall into this group, and the mechanism proposed for racemisation depends

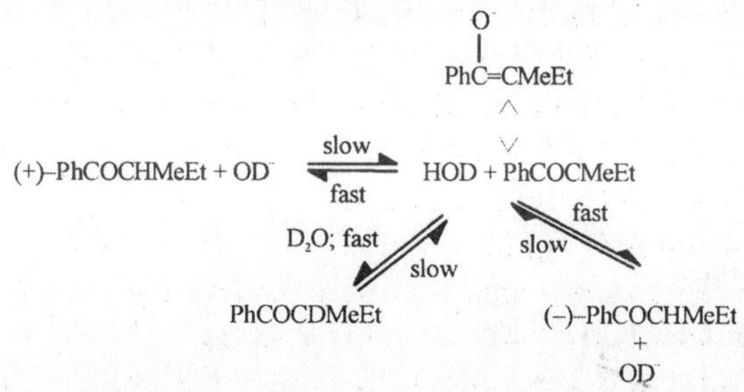

on the type of compound under consideration. In the case of compounds of the type of (–)-limonene which is racemised by strong heating, the mechanisms proposed are highly speculative, *e.g,* according to Kincaid *et al.,* (1940), molecules of the type Cabde can only be racemised by the breaking of bonds. A number of optically active secondary alcohols can be racemised by heating with a sodium alkoxide. This has been explained by a reversible dehydrogenation (Huckel, 1931) and there is some evidence to support this mechanism (Doering *et al.,* 1947, 1949). It has also been found that the presence of a trace of carbonyl compound (generally formed by atmospheric oxidation) is necessary for this reaction.

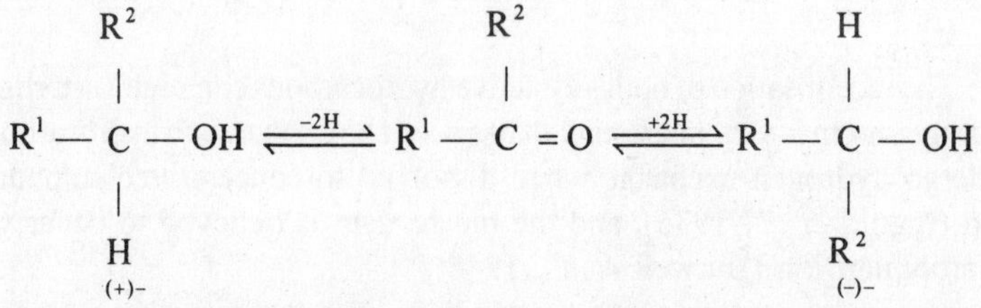

Another different type of compound which can be readily racemised is that represented by a-chloroethylbenzene. When the (+)- or (-)-form is dissolved in liquid sulphur dioxide, spontaneous racemisation occurs.

This has been explained by assuming ionisation into a carbonium ion (Polanyi *et al.,* 1933).

$$\underset{(+)\text{-}}{C_6H_2CHClCH_3} \rightleftharpoons C_6H_5\overset{+}{C}HCH_3 + Cl^- \rightleftharpoons \underset{(-)\text{-}}{C_6H_5CHClCH_3}$$

The carbonium ion is planar (the positively charged carbon atom is in a state of trigonal hybridisation) and consequently symmetric; recombination with the chlorine ion can occur equally well to form the (+)- and (–)-forms, *i.e.,* racemisation occurs. The basis of this mechanism is that alkyl halides in liquid sulphur dioxide exhibit an electrical conductivity, which has been taken as indicating ionisation. Hughes, Ingold *et al.,* (1936), however, found that pure a-chloroethylbenzene in pure liquid sulphur dioxide does not conduct, but when there is conduction, then styrene and hydrogen chloride are present. These authors showed that under the conditions of purity, the addition of bromine leads to a quantitative yield of styrene dibromide, and so suggested that the rate of racemisation is accounted for by the rate of formation of hydrogen chloride; thus:

$$C_6H_5CHClCH_3 \xrightarrow{\text{slow}} C_6H_5\overset{+}{C}HCH_3$$

$$C_6H_5\overset{+}{C}HCH_3 \xrightarrow{\text{fast}} C_6H_5CH = CH_2 + H^+$$

It is the recombination of the styrene with the hydrogen chloride that produces the racemised product; this may be written as follows

$$\underset{(+)\text{-}}{C_6H_5CHClCH_3} \rightleftharpoons C_6H_5CH = CH_2 + HCl \rightleftharpoons \underset{(-)\text{-}}{C_6H_5CHClCH_3}$$

α-Chloroethylbenzene can also be readily racemised by means of Lewis acids, *e.g,* SbC15, $HgC1_2$, etc. In this case, the mechanism is believed to be similar to that proposed by Polanyi (see above). Thus:

$$\underset{(+)\text{-}}{C_6H_5CHClCH_3} + SbC1_5 \rightleftharpoons C_6H_5\overset{+}{C}HCH_3 + SbCl_6^- \rightleftharpoons C_6H_5CHClCH_3 + SbCl_5$$

The racemisation of optically active hydrocarbons containing a tertiary hydrogen atom is very interesting. It has been shown that such hydrocarbons undergo hydrogen exchange when dissolved in concentrated sulphuric acid (Ingold *et al.,* 1936), and the mechanism is believed to occur via a carbonium ion (Burwell *et al.,* 1948).

$$R_3CH + 2H_2SO_4 \rightarrow R_3C^+ + HSO_4^- + SO_2 + 2H_2O$$

$$R_3C^+ + R_3CH \rightarrow R_3CH + R_3C^+, \text{ etc.}$$

This reaction is very useful for racemising optically active hydrocarbons, *e.g,* Burwell *et al.,* (1948) racemised optically active 3-methylheptane in concentrated sulphuric acid (the carbonium ion is flat):

$$\underset{(+)-}{C_2H_5-\overset{\overset{\displaystyle CH_3}{|}}{CH}}-C_4H_9+C_2H_5-\overset{\overset{\displaystyle CH_3}{|}}{C^+}-C_4H_9 \longrightarrow$$

$$C_2H_5-\overset{\overset{\displaystyle CH_3}{|}}{C^+}-C_4H_9+C_2H_5-\underset{(\pm)-}{\overset{\overset{\displaystyle CH_3}{|}}{CH}}-C_4H_9$$

Optically active hydrocarbons can also be racemised by means of aluminium chloride, the mechanism again probably being via the formation of a carbonium ion, *e.g,* 2-phenylbutane

$$\text{(I)}\quad \underset{(+)-}{C_6H_5CH(CH_3)C_2H_5}+AlCl_2 \longrightarrow C_6H_5\overset{+}{C}(CH_3)C_2H_5+HAlCl_2^-$$

$$\text{(II)}\quad \underset{(+)-}{C_6H_5CH(CH_3)C_2H_5}+C_6H_5\overset{+}{C}(CH_3)C_2H_5 \longrightarrow$$

$$C_6H_5\overset{+}{C}(CH_3)C_2H_5+\underset{(\pm)-}{C_6H_5CH(CH_3)C_2H_5}$$

The racemisation of other types of optically active compounds is described later.

PROPERTIES OF THE RACEMIC MODIFICATION

The racemic modification may exist in three different forms in the solid state.

(i) **Racemic mixture :** This is also .known as a (±)-conglomerate, and is a mechanical mixture of two types of crystals, the (+)- and (–)-forms; there are two phases present. The physical properties of the racemic mixture are mainly the same as those of its constituent enantiomers. The most important difference is the m.p.

(ii) **Racemic compound :** This consists of a pair of enantiomers in combination as a molecular compound; only one solid phase is present. The physical properties of a racemic compound are different from those of the constituent enantiomers, but in solution racemic compounds dissociate into the (+)- and (–)-forms.

(iii) **Racemic solid solution :** This is also known as a pseudo-racemic compound, and is a solid solution (one phase system) formed by a pair of enantiomers crystallising together due to their being isomorphous. The properties of the racemic solid solution are mainly the same as those of its constituent enantiomers; the m.p.s may differ.

Methods for Determining the Nature of a Racemic Modification

One simple method of examination is to estimate the amounts of water of crystallisation in the enantiomers (only one need be examined) and in the racemic modification; if these are different, then the racemic modification is a racemic compound. Another simple method is to measure the densities of the enantiomers and the racemic modification; again, if these are different, the recemic modification is a recemic compound; *e.g*, tartaric acids.

Table 1

	D-Tartaric acid	*L.-Tartaric acid*	*Racemic tartaric acid*
Melting point	170°C	170°C	206°C
Water of crystallisation	None	None	$1H_2O$
Density	1.7598	1.7598	1.697
Solubility in H_2O (at 20°C)	139 g/100 ml	139 g/100 ml	20.6 g/100 ml

There are, however, two main methods for determining the nature of a racemic modification: a study of the freezing-point curves and a study of the solubility curves (Roozeboom, 1899; Andriani, 1900).

Freezing Point Curves

These are obtained by measuring the melting points of mixtures containing different amounts of the racemic modification and its corresponding enantiomers. Various types of curves are possible according to the nature of the racemic modification. The melting points of all mixtures are higher than that of the racemic modification alone. In this case the racemic modification is a racemic mixture (a eutectic mixture is formed at the pcint of 50 per cent composition of each enantiomer), and so addition of either enantiomer to a racemic mixture raises the melting point of the latter; (±)-pinene is an example of this type. The melting points of the mixtures are lower than the melting point of the racemic modification which, therefore, is a racemic compound. When the racemic modification is a racemic solid solution, three types of curves are possible. In freezing-point curve is a horizontal straight line, all possible compositions having the same melting point, *e.g,* (+)-and (–)-camphor. The freezing-point curve shows a maximum, *e.g,* (+)- and (–)-carvoxime; and in the freezing-point curve shows a minimum, *e.g,* (+)- and (–)isopentyl (isoamyl) carbamate. In a number of cases there is a transition temperature at which one form of the racemic modification changes into another form, *e.g,* (±)-camphoroxime crystallises as the racemic solid solution above 103°C, whereas below this temperature it is the racemic compound that is obtained.

Correlation of Configurations by Means of Quasi-racemic Compounds

Fredga (1944) has introduced the study of quasi-racemic compounds as a means of correlating configurations their formation being detected by studying the melting-point curves of the two components. The curves obtained are similar to those of the racemic modification the quasi-racemic compounds these curves are unsymmetrical (since the m.p.s of the components will be different).

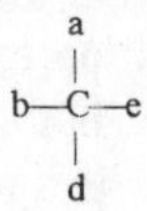

(I)

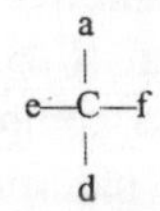

(II)

(I) and (II). On the other hand, curves of the other two types are given by compounds of like configuration (but some cases are known where the configurations have been opposite). Various examples of this method of correlating configurations have now been described; *e.g,* Fredga (1941) showed (partly by chemical methods and partly by using the quasi-racemate method) that (+)-malic acid (III) and (–)-mercaptosuccinic acid (IV) had opposite configurations. He then showed (1942) that

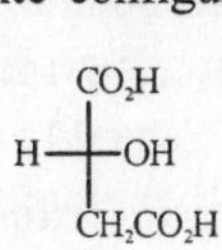

(III)

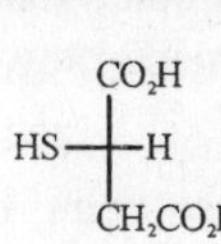

(IV)

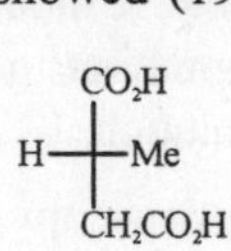

(V)

compound with (+)-methylsuccinic acid (V). Therefore (IV) and (V) have opposite configurations and consequently (+)-malic acid and (+)-methylsuccinic acid have the sarne configuration. It is of interest to note that McPhail *et al.,* (1966) have confirmed, by X-ray analysis, the absolute configuration of methylsuccinic acid established by Fredga.

Mislow *et al.,* (1956) have applied the m.p. curves in a somewhat different manner. They worked with 3-mercapto-octanedioic acid (VI) and 3-methyl-octanedioic acid (VII). These authors found that compounds (–)-VI and (+)-(VII) gave solid solutions for all mixtures (unsymmetrical 2.20a), whereas (+)-(VI) and (+)-(VII) give a diagram with a single

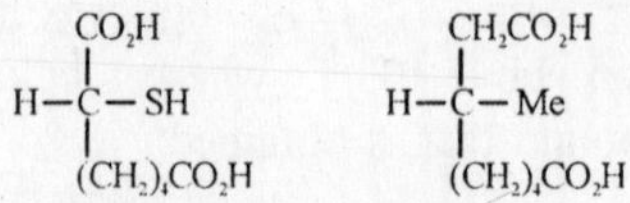

(–)-form (VI) (+)-form (VII)

eutectic (unsymmetrical 2.19a). These results indicate that (–)-(VI) and (+)-(VIII) have the same absolute configuration, whereas (+)-(VI) and (+)-(VII) have opposite configurations.

Solubility curves : The interpretation of solubility curves is difficult, but in practice the following simple scheme based on solubility may be used. A small amount of one of the enantiomers is added to a saturated solution of the racemic modification, and the resulting solution is then examined in a polarimeter. If the solution exhibits a rotation, then the racemic modification is a compound, but if the solution has a zero rotation, then the racemic modification is a mixture or a solid solution. The reasons for this behaviour are as follows.

If the racemic modification is a mixture or a solid solution, then the solution (in some solvent) is saturated with respect to each enantiomer and consequently cannot dissolve any of the added enantiomer. If, however, the racemic modification is a compound, then the solution (in a solvent) is saturated with respect to the compound form but not with respect to either enantiomer; hence the latter will dissolve when added and thereby produce a rotation. It should be noted that this simple method does not permit a differentiation to be made between a racemic mixture and a racemic solid solution.

Infrared spectroscopy is also being used to distinguish a racemic compound from a racemic mixture or a racemic solid solution. In the latter the spectra are identical, but are different in the former. These observations are also true for X-ray powder diagrams, and so X-ray analysis in the solid state may also be used.

RESOLUTION OF RACEMIC MODIFICATIONS

Resolution is the process whereby a racemic modification is separated into its two enantiomers. In practice the separation may be far from quantitative, and in some cases only one form may be obtained. Furthermore; the form isolated need not be optically pure, *i.e.,* it may consist of the (+)and (–)-forms in unequal amounts, but in this case the process is usually referred to as partial resolution. A large variety of methods for resolution have now been developed, and the method used in a particular case depends largely on the chemical nature of the compound under consideration.

(i) **Mechanical separation :** This method is also known as spontaneous resolution by crystallisation, and was introduced by Pasteur (1848). It depends on the crystallisation of the two forms separately, which are then separated by hand. The method is applicable only for racemic mixtures where the crystal forms of the enantiomers are themselves enantiomorphous. Pasteur separated sodium ammonium racemate in this way. The transition temperature of sodium ammonium racemate is 28°C; above this temperature the racemic compound crystallises out, and below this temperature the racemic mixture. Now Pasteur crystallised his sodium ammonium racemate from a concentrated solution at room temperature, which must have been below 28°C, since, had the temperature been above this, he would have obtained the racemic compound, which cannot be separated mechanically. Actually, Staedel (1878) failed to repeat Pasteur's separation since he worked at a temperature above 28°C.

(ii) **Preferential crystallisation by inoculation :** A supersaturated solution of the racemic modification is treated with a crystal of one enantiomer (or an isomorphous substance), whereupon this form is precipitated. The resolution of glutamic acid by inoculation has been perfected for industrial use (Ogawa *et al.,* 1957; Oeda, 1961). Harada *et al.,* (1962) have also resolved the copper complex of DL-aspartic acid by inoculation.

Except for the two amino-acids mentioned above, this method of resolution has been found impractical or resulted in partial resolution only. Harada (1965), however, has now obtained total optical resolution of free a-amino-acids by the inoculation method. Resolution was effected by seeding the supersaturated aqueous solutions with pure crystals of L- or D-isomer of the amino-acid.

(iii) **Biochemical separation (Pasteur, 1858) :** Certain bacteria and moulds, when they grow in a dilute solution of a racemic modification, destroy one enantiomer more rapidly than the other, *e.g,* Penicillium glaucum (a mould), when grown in a solution of racemic ammonium tartrate, attacks the (+)-form and leaves the (–).

This biochemical method of separation has some disadvantages

(a) Dilute solutions must be used, and so the amounts obtained will be small.

(b) One form is always destroyed and the other form is not always obtained in 50 per cent yield since some of this may also be destroyed.

(c) It is necessary to find a micro-organism which will attack only one of the enantiomers.

(iv) **Conversion into diastereoisomers (Pasteur, 1858) :** This method, which is the best of all the methods of resolution, consists in converting the enantiomers of a racemic modification into diastereoisomers the racemic modification is treated with an optically active substance and the diastereoisomers thereby produced are separated by fractional crystallisation. Thus, racemic acids may be separated by optically active bases, and vice versa, *e.g,*

$$(D_{acid} + L^{acid}) + 2D_{base} \rightarrow (D_{acid}D_{base}) + (L_{acid}D_{base})$$

These two diastereoisomers may then be separated by fractional crystallisation and the acids (enantiomers) regenerated by hydrolysis with inorganic acids or with alkalis. In practice it is usually easy to obtain the less-soluble isomer in a pure state, but it may be very difficult to obtain the moresoluble isomer.

In a number of cases this second (more-soluble) isomer may be obtained by preparing it in the form of another diastereoisomer which is less soluble than that of its enantiomer. On the other hand, separation and purification of the diastereoisomers may be successfully achieved by chromatography.

Resolution by means of diastereoisomer formation may be used for a variety of compounds, *e.g,*

(a) **Acids :** The optically active bases used are mainly alkaloids: brucine, quinine, strychnine, cinchonine, cinchonidine and morphine. Synthetic optically active bases are also used, *e.g,* benzimidazoles, menthylamine, α-phenylethylamine.

(b) **Bases :** Many optically active acids have been used, *e.g,* tartaric acid, camphor-β-sulphonic acid and particularly α-bromocamphor-π-sulphonic acid.

(c) **Alcohols :** These are converted into the acid ester derivative using either succinic or phthalic anhydride (Pickard and Kenyon, 1912). The acid ester, consisting of equimolecular amounts of

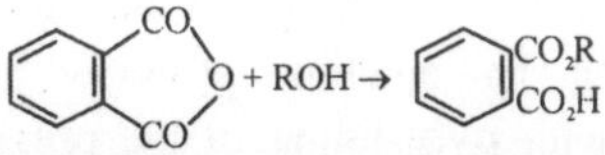

the (+)- and (-)-forms, may now be resolved as for acids. Racemic alcohols may also be resolved by diastereoisomer formation with optically active acyl chlorides (to form esters) or with optically active isocyanates (to form urethans)

$$R^1OCH_2COCl + R^2OH \rightarrow R^1OCH_2CO_2R^2 + HCl$$

$$R^1NCO + R^2OH \rightarrow R^2NHCO_2R^2$$

In these equations R^1 is the (–)-menthyl group; recently N-(–)-menthyl-p-sulphamylbenzoyl chloride, (I), has been used (Mills *et al.,* 1950).

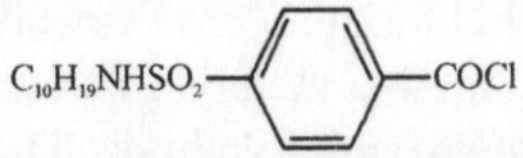

β-Acetoxy-D^5-etienic acid has been found very useful for resolving (±)-alcohols (inter alia, Djerassi *et al.,* 1961).

(d) **Aldehydes and Ketones :** These have been resolved by means of optically active hydrazines, *e.g,* (–)-menthylhydrazine. Sugars have been resolved with (+)-isopentanethiol. Nerdel *et al.,* (1952) have resolved oxo compounds with D-tartramide acid hydrazide,

$$NH_2COCHOHCHOHCONHNH_2;$$

this forms diastereoisomeric tartramazones. On the other hand, Shillington *et al.,* (1958) have converted oxo compounds into their 4-carboxyphenylsemicarbazones by means of a 4-carboxyphenylsemicarbazide,

Since these derivatives contain a carboxyl group, they can be resolved like acids, *e.g,* with brucine, and finally hydrolysed to liberate the optically active oxo compounds. Another method of resolution is reduction of the oxo compound to the corresponding alcohol, which is then resolved and the separated enantiomers re-oxidised.

Adams *et al.,* (1966) have resolved ketones via enamine formation. The enamine, produced by condensation of the ketone with pyrrolidine in the presence of a trace of p-toluenesulphonic acid is converted into iminium salts containing optically active anions, *e.g,* (+)-camphor10-sulphonate anion (represented as Z^- in the equation): Recrystallisation from suitable solvents gives the (+)- and (–)-forms.

(e) **Amino-compounds :** These may be resolved by conversion into diastereoisomeric anils by means of optically active aldehydes. Amines have also been resolved via their salts using, *e.g,* (+)-tartaric acid. a-Amino-acids have been resolved by preparing (he acyl derivative with an optically active acyl chloride, *e.g,* (–)-menthoxyacetyl chloride (cf. alcohols). Another method of resolving DL-amino-acids is asymmetric enzymic synthesis. The racemic amino-acid is converted into the acyl derivative which is then allowed to react with aniline in the presence of the enzyme papain at the proper pH (Albertson, 1951). Under these conditions only the L-amino-acid derivative reacts to form an insoluble anilide; the D-acid does not react but remains in the solution.

$$\underset{\text{DL-acid}}{R^1\overset{NHCOR^2}{\overset{|}{C}}HCO_2H} + C_6H_5NH_2 \xrightarrow{\text{papain}} \underset{\text{L-acid}}{R^1\overset{NHCOR^2}{\overset{|}{C}}HCONHC_6H_5} + \underset{\text{D-acid}}{R^1\overset{NHCOR^2}{\overset{|}{C}}HCO_2H}$$

Amino-acids have also been resolved by other means. Although amino-acids contain both an amino-group and a carboxyl group, they usually cannot be resolved in the straightforward way as amines or acids. This is due to the fact that amino-acids behave as dipolar ions.

Asymmetric Transformation

Resolution of racemic modifications by means of salt formation (the diastereoisomers are salts; cf. acids and bases) may be complicated by the phenomenon of asymmetric transformation. This is exhibited by compounds that are optically unstable, *i.e.,* the enantiomers are readily interconvertible

$$(+)\text{–C} \rightleftharpoons (-)\text{–C}$$

Suppose we have an optically stable (+)-base (one equivalent) dissolved in some solvent, and this is then treated with one equivalent of an optically unstable (±)-acid. At the moment of mixing, the solution will contain equal amounts of [(+)-Base .(+)-Acid] and [(+)-Base.(–)-Acid]; but since the acid is optically unstable, the two diastereoisomers will be present in unequal amounts when equilibrium is attained.

$$[(+)\text{-Baseo}(+)\text{-Acid}] \rightleftharpoons [(+)\text{-Base.}(-)\text{-Acid}]$$

According to Jamison and Turner, first-order asymmetric transformation is the establishment of equilibrium in solution between the two diastereoisomers which must have a real existence. In secondorder asymmetric transformation it is necessary that one salt should crystallise from solution; the two diastereoisomers need not have a real existence in solution. In second-order asymmetric transformation it is possible to get a complete conversion of the acid into the form that crystallises; the form may be the (+)- or (–)-, and which one it is depends on the nature of the base and the solvent.

Many examples of first- and second-order asymmetric transformation are known, and a large number of these compounds are those which owe their chirality to restricted rotation about a single bond, *e.g,* Mills and Elliott (1928) tried to resolve N-benzenesulphonyl-8-nitro-l-naphthylglycine, (II), by means of the brucine salt. These authors found that either diastereoisomer could be obtained in approximately 100 percent yield by crystallisation from methanol and acetone, respectively.

Another example of second-order asymmetric transformation is hydrocarbostyril-3carboxylic acid. This compound contains an asymmetric carbon atom, and Leuchs (1921), attempting to resolve it with quinidine, isolated approximately 90 per cent of the (+)-form. Optical instability in this case is due to keto-enol tautomerism. A very interesting example of second-order asymmetric transfor-mation is 2-acetomethylamido4',5-dimethylphenylsulphone, (III). When this compound was crystallised from a supersaturated solution in ethyl (+)-tartrate, the crystals obtained had a rotation of +02°; evaporation of the mother liquor gave crystals with a rotation of -0 15° (Buchanan *et al.,* 1950).

(v) Another method of resolution that has been tried is the conversion of the enantiomers into volatile diastereoisomers, which are then separated by fractional distillation. So far, the method does not appear to be very successful, only a partial resolution being the result; *e.g,* Bailey and Hass (1941) converted (±)-pentan-2-ol into its diastereoisomers with L(+)-lactic acid, and then partially separated them by fractional distillation.

(vi) **Chromatography** : Optically active substances may be selectively adsorbed by some optically active adsorbent, *e.g,* Henderson and Rule (1939) partially resolved p-phenylenebisiminocamphor on lactose as adsorbent; Bradley and Easty (1951) have found that wool and casein selectively adsorb (+)-mandelic acid from an aqueous solution of (±)-mandelic acid. A particularly important case of resolution by chromatography is that of Troger's base. Jamison and Turner (1942) have carried out a chromatographic separation without using an optically active adsorbent; they partially resolved the diastereoisomers of (–)-menthyl (±)mandelate by preferential adsorption on alumina. It is also interesting to note that the resolution of a racemic acid by salt formation with an optically active base is made more effective by the application of chromatography. More recently, enzymic and chromatographic methods have been developed for the direct separation of enantiomers (inter alia, Rogożhin, 1971; Gil-Av, 1972).

GSC and GLC have been used with great success for resolving racemic modifications, *e.g,* s-butanol and s-butyl bromide have been separated into two overlapping fractions using a column of starch or ethyl tartrate as the stationary phase (Karagounis *et al.,* 1959). On the other hand, Casanova *et al.,* (1961) have resolved the diastereoisomeric ketals from (±)-camphor by GLC, and Halpern *et al.,* (1965) have resolved in the same way DL-amino-acids via their (–)-menthyl ester derivatives and via their acyl derivatives. Amines may also be resolved via acylation with optically active acid chlorides. Halpern *et al.,* (1966) have used N-trifluoroacetyl-L-prolyl chloride, and separated the diastereoisomers by GLC, *e.g,* 2-aminobutane and 2-aminopentane have been obtained optically pure.

Beckett *et al.,* (1957) have introduced a novel method for correlating and determining configurations. These authors have prepared 'stereoselective adsorbents'. These are adsorbents prepared in the presence of a suitable reference compound of known configuration, *e.g,* silica gel in the presence of quinine. Such an adsorbent exhibits higher adsorptive power for isomers related to the reference compound than for their stereoisomers, provided that their structures are not too dissimilar from that of the reference compound. Thus, silica gel prepared in the presence of quinine adsorbs quinine more readily than its stereoisomer quinidine; cinchonidine (configurationally related to quinine) is adsorbed more readily than its stereoisomer cinchonine (configurationally related to quinidine).

(vii) **Kinetic method of resolution :** Marckwald and McKenzie (1899) found that (–)-menthol reacts more slowly with (–)-mandelic acid than with the (+)-acid. Hence, if insufficient (–)-menthol is used to completely esterify (±)-mandelic acid, the resulting mixture of diastereoisomers will contain more (–)-menthyl (+)-mandelate than (–)-menthyl (–)-mandelate. Consequently there will be more (–)-mandelic acid than (+)-mandelic acid in the unchanged acid, *i.e.,* a partial resolution of (±)-mandelic acid has been effected.

(viii) Ferreira (1953) has partially, resolved (±)-narcotine and (±)-laudanosine (1–2.5 per cent resolution) without the use of optically active reagents. He dissolved the racemic alkaloid in hydrochloric acid and then slowly added pyridine; the alkaloid was precipitated, and it was found to be optically active. The explanation offered for this partial resolution is as follows (Ferreira). When a crystalline racemic substance is precipitated from solution, a crystallisation nucleus is first developed. Since this nucleus contains a relatively small number of molecules, there is more than an even chance that it will contain an excess of one enantiomer or other. If it be assumed that the forces acting on the growth of crystals are the same kind as those responsible for adsorption [cf. (vi)], the nucleus will grow preferentially, collecting one enantiomer rather than the other. Crystallisation, when carried out in the usual manner, results in the formation of crystals containing more or less equivalent numbers of both enantiomers.

(ix) **Channel complex formation** has also been used to resolve racemic modifications. This also offers a means of carrying out a resolution without chiral reagents, *e.g,* Schlenk (1952) added (±)-2-chloro-octane to a solution of urea and obtained, on fractional crystallisation, the two urea inclusion complexes urea/(+)-2-chloro-octane and urea/(–)-2-chloro-octane.

Baker *et al.,* (1952) have prepared tri-o-thymotide, and found that it formed clathrates with ethanol, n-hexane, etc. Powell *et al.,* (1952) have shown that tri-o-thymotide crystallises as a racemate, but that resolution takes place when it forms clathrates with n-hexane, benzene or chloroform. By means of seeding and slow growth of a single crystal, it is possible to obtain the (+)- or (–)-form depending on the nature of the seed. Furthermore, crystallisation of tri-o-thymotide (dl) from a solvent which is itself a racemic modification (d′l′) and which forms a clathrate, produces crystals of the types dd′ and ll′. Thus such (solvent) racemic modifications can be resolved, *e.g,* s-butyl bromide has been resolved in this way.

Optical Purity

An optically pure compound is one which has been prepared in 100 per cent purity, *i.e.,* optical purity is expressed as a percentage, *e.g,* if the (maximum) specific rotation of compound A is + 50° and an impure sample has a rotation of + 30°, this sample is 60 per cent optically pure. For a racemic modification, the optical purity is zero.

The difficult problem with respect to optical purity is to be able to ascertain whether a specimen of an enantiomer is 100 per cent optically pure. Several criteria may be used. The simplest criterion is that which considers a crystalline compound to be optically pure if, after repeated crystallisation, the melting point and rotation remain unchanged. This, however, may not be a correct conclusion, *e.g,* the resolution of a racemic solid solution may lead to the isolation of a partially resolved enantiomer which, after repeated crystallisation, does not change its rotation. The conclusion, however, that the compound is optically pure is strengthened if both of its enantiomers can be prepared and their rotations are equal and opposite.

There are several methods which may be used for ascertaining optical purity and are reliable within certain experimental limits. One method uses isotopic dilution. A known weight of the enantiomer being examined is mixed, in solution, with a known weight of its racemic modification which has been labelled with an isotope. After recrystallisation, the isotope content of the racemic modification is then determined. Suppose the enantiomer under consideration is the (+)-form. In this case, the recovered racemic modification will contain unlabelled (+)-form as well as labelled (+)-form, and it is therefore possible to calculate the dilution factor (since known weights of both were used). If, however, the (+)-enantiomer is not optically pure, some unlabelled (–)-form will also be present in the recovered racemic modification. In this case, the isotope dilution factor will be less than the predicted one.

Other methods make use of enzymes or conversion into other compounds of known optical purity. NMR spectroscopy may also be used to determine optical purity. It has already been pointed out that the NMR spectra of a pair of enantiomers are identical. Now suppose that a racemic modification is completely converted into a pair of diastereoisomers, *e.g,*

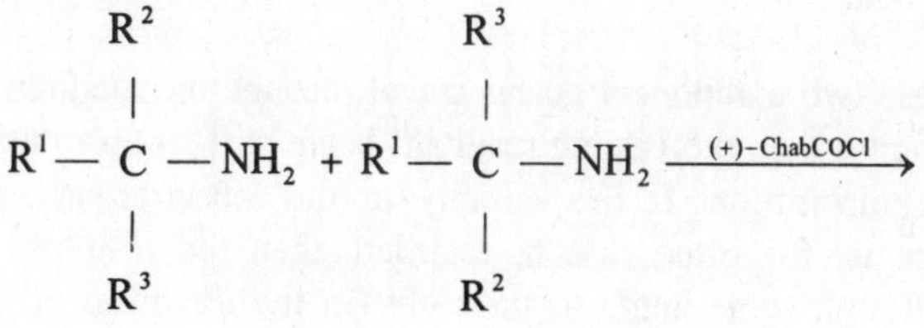

(+)-(I) (–)-(I)

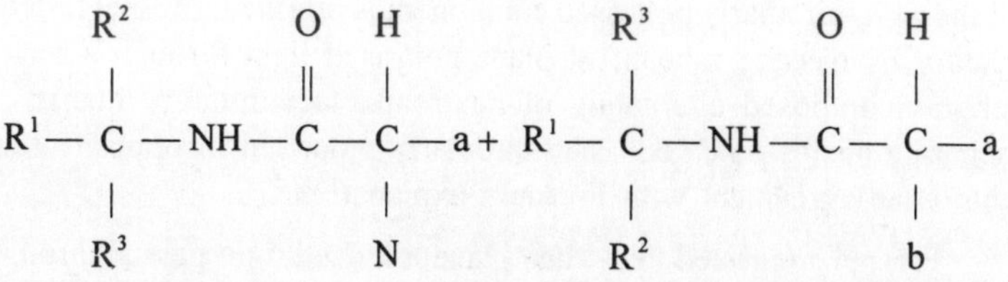

(IIa) (IIb)

In the enantiomers (+)-(I) and (–)-(I), corresponding pairs of groups are enantiotopic and their chemical shifts are identical. This is no longer the case for corresponding groups in (IIa) and (IIb). Corresponding pairs are now in diastereoisomeric environments. Thus, the protons of the group CH are diastereotopic, and if the acid chloride is optically pure, the two different proton signals will have the same intensities in their NMR spectra. If reaction between the optically pure acid chloride (in excess) is carried out with a resolved specimen of (I), then only one signal for the CH proton will be observed if (I) is optically pure. If (I) is not optically pure, then (IIa) and (IIb) will be formed in unequal amounts and two signals will be observed for the CH proton. It is then possible to calculate the optical purity of (1) from the ratio of the intensities of the two signals.

THE CAUSE OF OPTICAL ACTIVITY

Two important points that arise from the property of optical activity are: What types of structure give rise to optical activity, and why? Fresnel (1822) suggested the following explanation for optical activity in crystalline substances such as quartz, basing it on the principle that any simple harmonic motion along a straight line may be considered as the resultant of two opposite circular motions. Fresnel assumed that plane-polarised light, on entering a substance in a direction parallel to its optic axis, is resolved into two beams of circularly polarised light, one right-handed (dextro-) and the other left-handed (laevo-) and both having the same frequency.

If these two component beams travel through the medium with the same velocity, then the issuing resultant beam suffers no rotation of its plane of polarisation. If the velocity of the left-circularly polarised component is, for some reason, retarded, then the resultant beam is rotated through some angle to the right (in the direction of the faster circular component. Similarly, the resultant beam is rotated to the left if the right-circularly polarised component is retarded. Fresnel tested this theory by passing a beam of plane-polarised light through a series of prisms composed alternately of dextroand laevorotatory quartz. Two separate beams emerged, each circularly polarised in opposite senses; this is an agreement with Fresnel's explanation.

Fresnel suggested that when planepolarised light passed through an optically active crystalline substance, the plane of polarisation was rotated because of the retardation of one of the circular components. Stated in another way, Fresnel's theory requires that the refractive indices for right- and left-circularly polarised light should be different for optically active substances. It has been shown mathematically that only a very small difference between these refractive indices gives rise to fairly large rotations, and that if the refractive index for the left-circularly polarised light is greater than that for the right component the substance, will be dextrorotatory. The difficulty of Fresnel's theory is that it does not explain why the two circular components should travel with different velocities. It is interesting to note, however, that Fresnel (1824) suggested that the optical activity of quartz is due to the structure being built up in right- and left-handed spirals.

If n_L and n_R respectively represent the refractive indices for left- and right-circularly polarised light then, if these are different, the substance is said to exhibit circular birefringence. If $n_L > n_R$, the emergent linearly-polarised resultant is rotated to the right. Since refraction and absorption of light are interconnected, the implication is that if $n_L > n_R$ in the long wavelength region where the optically active substance is transparent, then $\varepsilon_L > \varepsilon_R$ (where ε is the molar absorptivity) for the shorter wavelength region where light is absorbed. This effect, *i.e.,* when ε_L and ε_R are unequal, is known as circular dichroism.

The combined phenomenon of circular dichroism and unequal velocity of travel of left- and right-circularly polarised light is known as the Cotton effect. Now, ORD and CD curves are studied in the region of maximum absorption of the optically active compound, i..e., in the region of an optically active chromophore. Such a chromophore is either inherently asymmetric, *e.g,* twisted biphenyls, or inherently symmetric, *e.g,* a carbonyl group. In the latter, when this is in an asymmetric environment, it then behaves as an asymmetric chromophore, *i.e.,* optical activity is induced in the chromophore by the environment. Because of this, the carbonyl chromophore is referred to as an inherently symmetric, but asymmetrically perturbed, chromophore. The amplitudes of the ORD curves of compounds containing inherently asymmetric chromophores are usually much greater than those containing asymmetrically perturbed symmetric chromophores.

Drude (1900) showed that if a molecule possessed a structure such that when light is absorbed an electron is displaced along a right-handed helical path, then the result is a positive circular dichroism ($\varepsilon_L > \varepsilon_R$), and the molecule is dextrorotatory at longer wavelengths where $n_L > n_R$. In the enantiomer, the electron is displaced along a left-handed helical path when light is absorbed, the result being a negative circular dichroism ($\varepsilon_R > \varepsilon_L$) and an optical laevorotation ($n_R > n_L$) at longer wavelengths.

This theory of optical rotatory power has been modified by quantum mechanics treatment. A helical motion is the resultant of two components, a linear and a circular displacement. A linear and a circular charge-displacement produce an electric and magnetic dipole moment, respectively. If a transient electric dipole and a magnetic dipole are produced by absorption of light, the molecule is circularly dichroic. If the two moments are parallel, the circular dichroism is-positive; if the two moments are antiparallel, the circular dichroism is negative. If the two moments are mutually perpendicular, then no circular dichroism results.

Now let us consider the problem of optical activity of substances in solution. In this case the optical activity is due to the molecules themselves, and not to crystalline structure. Any crystal which has a plane of symmetry but not a centre of symmetry rotates the plane of polarisation, the rotation varying with the direction in which the light travels through the crystal. No rotation occurs if the direction of the light is perpendicular or parallel to the plane of symmetry.

If we assume that molecules in a solution (or in a pure liquid) behave as individual crystals, then any molecule having a plane but not a centre of symmetry will also rotate the plane of polarisation, provided that the light travels through the molecule in any direction other than perpendicular (or parallel) to the plane of symmetry.

Let us consider the molecule Ca_2bd. This has a plane of symmetry, and so molecule (I) and its mirror image (II) are superimposable. Now let us suppose that the direction of plane-polarised light passing through molecule (I) makes an angle 0° with the plane of symmetry, and that the resultant rotation is $+\alpha°$. Then if the direction of the light through molecule (II) also makes an angle $\theta°$ with the plane of symmetry, the resultant rotation will be $-\alpha°$. Thus the total rotation produced by molecules (I) and (II) is zero. In a solution of compound Ca_2bd there will be an infinite number of molecules in random orientation.

Statistically one can expect to find that whatever the angle θ is for molecule (I), there will always be molecule (II) also being traversed by light entering at angle θ. Thus, although each individual molecule rotates the plane of polarisation by an amount depending on the value of θ, the statistical sum of the contributions of the individual molecules will be zero.

When a molecule is not superimposable on its mirror image, then if only one enantiomer is present in the solution, the rotation produced by each individual molecule will (presumably) depend on the angle of incidence (with respect to any face), but there will be no compensating molecules (*i.e.,* mirror image molecules) present. Hence, in this case, there will be a net rotation that is not zero, the actual value being the statistical sum of the individual contributions (which are all in the same direction).

Thus, if we consider the behaviour of a compound in a solution (or as a pure liquid) as a whole, then the observed experimental results are always in accord with the statement that if the molecular structure of the compound is chiral, that compound will be optically active. Any compound composed of molecules possessing a plane but not a centre of symmetry is, considered as a whole, optically inactive, the net zero rotation being the result of 'external compensation'. This point is of great interest in connection with molecules that can exist in different conformations.

Let us consider meso-tartaric acid, a compound that is optically inactive by internal compensation. X-ray studies (Stern *et al.,* 1950) have shown that the staggered form of the molecule is the favoured one. This has a centre of symmetry, and so molecules in this configuration are individually optically inactive. On the other hand, meso-tartaric acid is usually represented by the plane-diagram formula. This corresponds to the eclipsed form, and has a plane of symmetry. In this conformation the individual molecules are optically active except when the direction of the light is perpendicular (or parallel) to the plane of symmetry; the net rotation is zero by 'external compensation'. It is possible, however, for the molecule to assume, at least theoretically, many conformations which have no elements of symmetry, *e.g.*

All molecules in this conformation will contribute in the same direction to the net rotation. If the total number of molecules present were in this conformation, then meso-tartaric acid would have some definite rotation. On the theory of probability, however, for every molecule taking up the conformation in Fig. 2.39(c). there will also be present its mirror image molecule, thereby giving a net zero rotation due to 'external compensation'. As we have seen, meso-tartaric acid is optically inactive (as shown experimentally), and by common usage the inactivity is said to be due to internal compensation.

NATURE OF GEOMETRICAL ISOMERISM

In this chapter we will study a new type of isomerism, where isomers possess the same structural formula containing a double bond and differ only in respect of the arrangement of atoms or groups about the double bond. Maleic and fumaric acids both have the same molecular formula $C_4H_4O_4$, but differ in most of their physical and in many of their chemical properties, and neither is optically active.

It was originally thought that these two acids were structural isomers; this is the reason for different names being assigned to each form (and to many other geometrical isomers). It was subsequently shown, however, that maleic and fumaric acids were not structural isomers, *e.g,* both (i) are catalytically reduced to succinic acid; (ii) add one molecule of hydrogen bromide to form bromosuccinic acid; (iii) add one molecule of water to form malic acid; (iv) are oxidised by alkaline potassium permanganate to tartaric acid (the stereochemical relationships in reactions (ii), (iii) and (iv) have been ignored.

Thus, both acids have the same structure, viz., $HO_2CCH= CHCO_2H$. van't Hoff (1874) suggested that if we assume there is no free rotation about a double bond, two spatial arrangements are possible for the formula $HO_2CCH=CHCO_2H$, and these would account for the isomerism exhibited by maleic and fumaric acids. Using tetrahedral diagrams, van't Hoff represented a double bond by placing the tetrahedra edge to edge. From a mechanical point of view, such an arrangement would be rigid, *i.e.,* free rotation about the double bond is not to be expected.

DEFINITION OF GEOMETRICAL ISOMERISM

According to Le Bel and van't Hoff (1874), the valencies of the carbon atom are distributed symmetrically in space. They may be pictured as directed towards the four corners of a regular tetrahedron constructed round the carbon atom as the centre. According to this theory, a derivative of ethane, say succinic acid, containing two carbon atoms united by a single bond. If the two trahedra were fixed rigidly to each other, the two compounds having the models I and II should be possible. Actually there is only one succinic acid known. The natural conclusion is that the two carbon atoms are able to rotate about the axis formed by the bond between them and can, therefore, assume form I or II.

If, however, the two carbon atoms arc united by a double bond, the two tetrahedra representing them must necessarily be in contact at two corners. In such a case all free rotation of the tetrahedra ceases and the groups attached to the two carbon atoms are fixed relative to each other. Thus, different spatial arrangements of the groups about the doubly bonded carbon atoms now become possible. For rahcdral models representing maleic acid and fumaric acid will be as shown

For the sacke of convenience the above models are represented in a plane as follows :

maleic acid (*cis*)	fumaric acid (*trans*)
H—C—COOH ‖ H—C—COOH	H—C—COOH ‖ HOOC—C—H

It may be noted clearly that meleic acid and fumaric acid have quite different arrangements of groups about the double bond. While in one case similar groups are on the same side, in the other they lie on opposite sides. *This type of isomerism due to the different geometrical arrangements of the groups about the doubly bonded carbon atoms* is known as *Geometrical isomerism.* When the similar groups lie on tht same side as in maleic acid, it is called the *Cis isomer* (Latin, *cis* = on same side). When similar groups are on the opposite sides as in the fumaric acid, it is called the *Trans isomer* (Latin, *trans* = across). Consequently this type of isomerism is often spoken of as *Cis-Trans isomerism.*

3

Optical Inversion (Walden Inversion)

INTRODUCTION

The series of replacement reactions, Walden (1893, 1895) transformed an optically active compound into its enantiomer. In some cases the product is 100 per cent optically pure, *i.e.*, the inversion is quantitative; in other cases the product is a mixture of the (+)- and (−)-forms in unequal amounts, *i.e.*, inversion and retention (racemisation) have taken place.

The phenomenon was first discovered by Walden with the following reactions

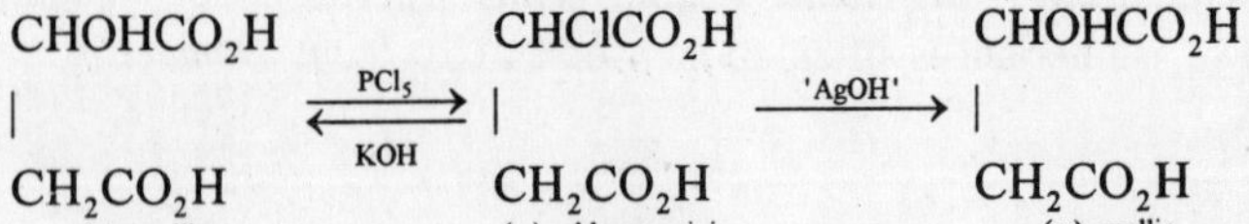

(−)-mallic acid (I) (+)-chlorosuccinic acid (II) (+)-mallic acid (III)

In one, and only one, of the two reactions must there be an interchange of position between the two groups, *e.g.*, if the configuration of (I) corresponds with that of (II), the inversion of configuration must have taken place between (II) and (III). The term Walden inversion is applied to any step in a reaction in which inversion of configuration occurs.

As the above experiment stands, there is no way of telling which step is accompanied by inversion. As we have seen, change in sign of rotation does not necessarily mean that inversion of configuration has occurred. Various methods of correlating configuration have already been described, but here we shall describe the method where bonds attached to the chiral centre are broken during the course of the reactions. This method was established by Kenyon *et al.* (1925), who carried out a series of reactions on optically active hydroxy compounds. Now it has been established that in the esterification of a monocarboxylic acid by an alcohol under ordinary conditions, the reaction proceeds by the acyl-oxygen fission mechanism thus;

$$R^1CO—OH + H—OR^2 \rightarrow R^1COOR^2 + H_2O$$

Kenyon assumed that in all reactions of this type the R^2—O bond remained intact and consequently no inversion of the alcohol is possible. The following chart shows a series of reactions carried out on ethyl (+)-lactate; Ts = tosyl group = p-toluenesulphonyl group, p-$McC_6H_4SO_2$-; the symbol → is used to represent inversion of configuration in that step. (IV) and (VI) have the same relative configurations even though the sign of rotation has changed. Similarly, (IV) and (V) have the same relative configurations. Reaction of (V) with potassium acetate, however, produces (VII), the enantiomer of (VI).

Therefore inversion must have occurred in the formation of (VII); (V) and (VI) are produced without inversion since in these cases the C–O bond in (IV) is never broken. It should be noted here that if inversion is going to take place at all, the complete group attached to the chiral centre must be removed (in a displacement reaction). The converse, however, is not true, *i.e.*, removal of a complete group does not invariably result in inversion.

The above series of reactions has been used as a standard, and all closely analogous reactions are assumed to behave in a similar way, *e.g.*, the action of lithium chloride on the tosylate (V) is assumed

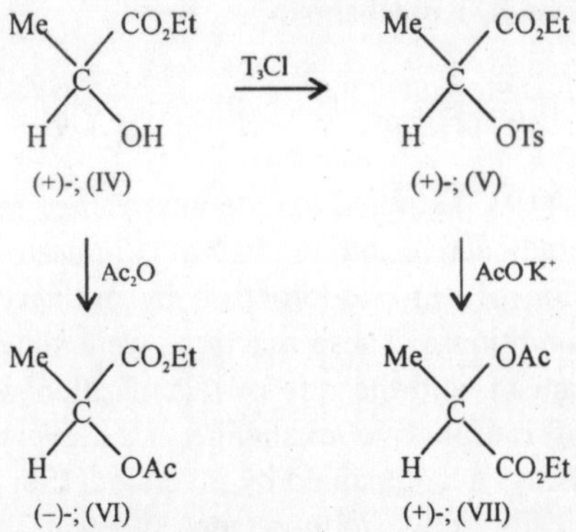

to be analogous to that of potassium acetate, and the chloride produced thus has an inverted configuration

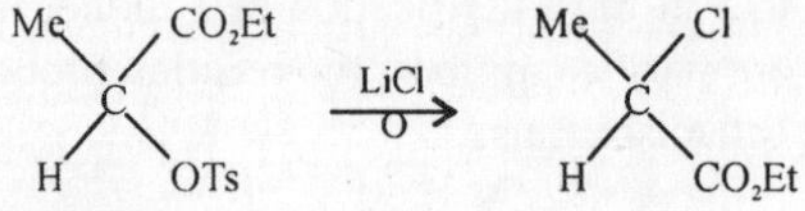

(V)

By similar procedures, Kenyon *et al.* (1929, 1930) showed that (+)-octan-2-ol and (+)-2-chloro-, 2-bromo- and 2-iodo-octane have the same relative configurations; and also that (+)-α-hydroxyethylbenzene (PhCHOHMe), (+)-a-chloro- and (+)-a-bromoethylbenzene have the same relative configurations.

WALDEN INVERSION MECHANISM

As the result of a large amount of work on the Walden inversion, it has been found that at least three factors play a part in deciding whether inversion or retention (racemisation) will occur:

(i) the nature of the reagent;

(ii) the nature of the substrate;

(iii) the nature of the solvent.

Hence, it is necessary to explain these factors when dealing with the mechanism of the Walden inversion.

Many theories have been proposed, but we shall discuss only the Hughes-Ingold theory, since this is the one now accepted. According to this theory, aliphatic nucleophilic substitution reactions may take place by either the S_N2 or S_N1 mechanism.

S_N2

$$HO^- \; R—X \rightarrow \overset{\delta-}{H}O\text{---}R\text{---}\overset{\delta-}{X} \rightarrow HO—R + X^-$$

Hughes *et al.* (1935) studied (a) the interchange reaction of (+)-2-iodo-octane with radioactive iodine (as NaI*) in acetone solution, and (b) the racemisation of (+)-2-iodo-octane by ordinary sodium iodide under the same conditions. These reactions were shown to take place by the S_N2 mechanism, and the rate of racemisation was shown to be twice the rate of radioactive exchange, *i.e.*, every iodide-iodide displacement is always accompanied by inversion. (Suppose there are n molecules of optically active iodo-octane. When n/2 molecules have exchange with I and in doing so have been inverted, racemisation is now complete although the exchange has taken place with only half of the total number of molecules.)

Thus, this experiment leads to the assumption that inversion always occurs in the S_N2 mechanism. This is fully supported by other experimental work. *e.g.*, Hughes *et al.* (1936, 1938) studied the reaction of optically active a-bromoethylbenzene and a-bromopropionic acid with radioactive bromide ions, and again found that the rate of racemisation was twice the rate of exchange.

Since S_N2 reactions always occur with inversion (this is known as the stereokinetic rule for S_N2 reactions), then provided the molecularity of the S_N reaction can be determined kinetically, it is possible to correlate the configuration of the reactant with that of the product. There are four S_N2 charge-types of reaction

		Reagent	***Substrate***
1.	$Y^- + RX \rightarrow YR + X^-$	negative	neutral
2.	$Y^- + RX^+ \rightarrow YR + X$	negative	positive
3.	$Y + RX \rightarrow YR^+ + X^-$	neutral	neutral
4.	$Y + RX^+ \rightarrow YR^+ + X$	neutral	positive

The stereokinetic rule for S_N2 reactions is well established for only reactions of type 1. Hughes, Ingold *et al.* (1960) have also shown that the rule applies to type 2, *e.g.*, the reaction between a sulphonium iodide and sodium azide:

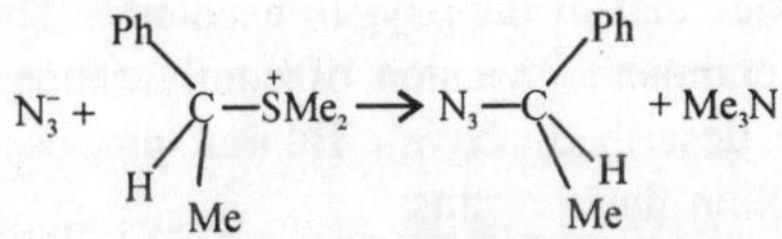

Another example is the reaction between the acetoxyl ion and the (+)-trimethyl-x-phenylethylammonium ion to give the inverted product (Snyder *et al.*, 1949):

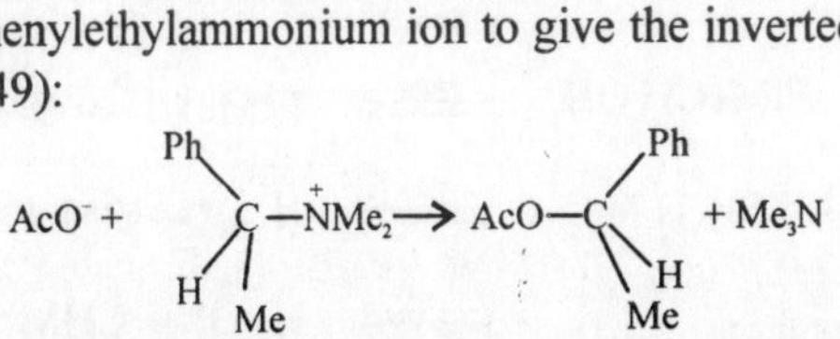

Hughes *et al.* (1964) have also shown that the reaction between thiourea and the dimethyl-lphenylsulphonium ion (as iodide) in methyl cyanide solution gives the thiouronium salt with substantially complete inversion; this is an example of type 4.

$$SC(NH_2)_2 + CHMePhSMe_2^+ \rightarrow CHMePhSC(NH_2)_2^+ + Me_2S \quad ...(1)$$

Examples of type 3 are also known, *e.g.*, the Menschutkin reaction.

$$R_3^1N + R^2—Br \rightarrow R_3^1\overset{+}{N}—R^2Br^-$$

Now let us consider the S_N1 mechanism.

$$R—N \rightarrow \overset{\delta+}{R}—\overset{\delta-}{X} \rightarrow R^+ + N^- \xrightarrow{OH^-} ROH + X^-$$

When the reaction proceeds by this mechanism, then inversion and retention (racemisation) will occur, the amount of each depending on various factors. The carbonium ion is flat (trigonal hybridisation), and hence attack by the nucleophilic reagent can take place equally well on either side, *i.e.*, equal amounts of the (+)- and (–)-forms will be produced; this is racemisation. Furthermore, on the basis that an ion-pair is formed first, the leaving group will protect its side from attack by the nucleophile, *i.e.*, inversion will occur exclusively or will predominate. Only if there is complete dissociation of the ion-pair into the individual ions can complete racemisation be expected; the shielding effect of the retiring ion is now lost.

Also, racemisation will be encouraged by low concentration of the nucleophile. An example of inversion is that due to Bunton *et al.* (1955), who studied the reaction of ^{18}O-enriched water with optically active s-butanol in aqueous perchloric acid, and found that the overall rate of racemisation is twice that of the oxygen exchange. Thus every oxygen exchange causes complete inversion of configuration (cf. the iodide-iodide* exchange described above). Bunton proposed the following mechanism to explain these results:

$$EtMeCHOH + H^+ \underset{fast}{\overset{HClO_4}{\rightleftharpoons}} EtMeCHOH_2^+ \quad (2)$$

$$EtMeCHOH_2^+ \overset{slow}{\rightleftharpoons} EtMe\overset{\delta+}{C}H---\overset{\delta+}{O}H_2 \quad (3)$$

$$H_2O^* + EtMe\overset{\delta+}{C}H---\overset{\delta+}{O}H_2 \overset{fast}{\rightleftharpoons} \overset{\delta+}{H_2O^*}---EtMeCH---\overset{\delta+}{O}H_2$$

$$\overset{fast}{\rightleftharpoons} H_2\overset{+}{O}^*—CHMeEt + H_2O \quad (4)$$

$$H_2\overset{+}{O}^*—CHMeEt \overset{fast}{\rightleftharpoons} HO^*—CHMeEt + H^* \quad (5)$$

(4) occurs before the OH_2^+ has completely separated in (3), and so this side is shielded and the H_2O^* is forced to attack on the other side as shown; the result is thus inversion. The above reaction proceeds by the

S_N1 mechanism since (3) is the rate-determining step (only one molecule is undergoing covalency change in this step). Had the reaction been S_N2, complete inversion would have been obtained. It was shown, however, that the reaction rate was independent of the concentration of H_2O^*. The mechanism is therefore SN 1, since had it been S_N2, the kinetic expression would require the concentration of the H_2O^*:

$$H_2O^* + EtMeCHOH_2^+ \xrightleftharpoons{\text{slow}} \overset{\delta+}{H_2O^*}\text{---}EtMeCH\text{---}\overset{\delta+}{H_2O^*}$$

$$\xrightarrow{\text{fast}} H_2\overset{+}{O^*}\text{—}CHMeEt + H_2O$$

In general, net inversion is usually observed for short-life carbonium ions. Ions of this type are produced from s- and t-alkyl derivatives. On the other hand, diarylmethyl carbonium ions have a long life (due to spreading of charge).

Thus, Ar^1Ar^2CHX undergoes solvolysis by the S_N1 mechanism to give a completely racemised product. These observations offer a means of estimating the relative stabilities of carbonium ions. The stereochemical course of S_N1 reactions may also be affected by neighbouring group participation.

THE S_Ni MECHANISM

Another important S_N reaction is the S_Ni type (substitution, nucleophilic, internal). The reaction between thionyl chloride and alcohols has been studied extensively. A well-examined example is the alcohol a-phenylethanol, PhCHOHMe; this is an arylmethanol, and according to Hughes, Ingold *et al.* (1937) the first step is the formation of a chlorosulphite. No inversion occurs at this stage (which is a four-centre reaction); in the following equation, R = PhMeCH-.

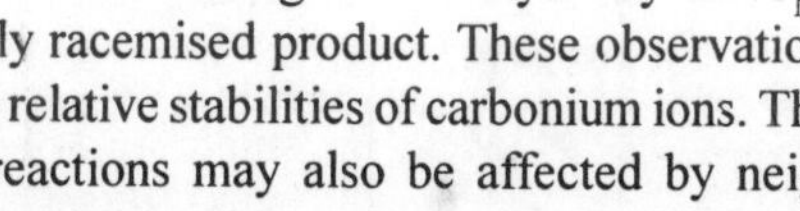

This chlorosulphite could then form a-chloroethylbenzene by one or more of the following mechanisms

(i) S_N2. This occurs with inversion.

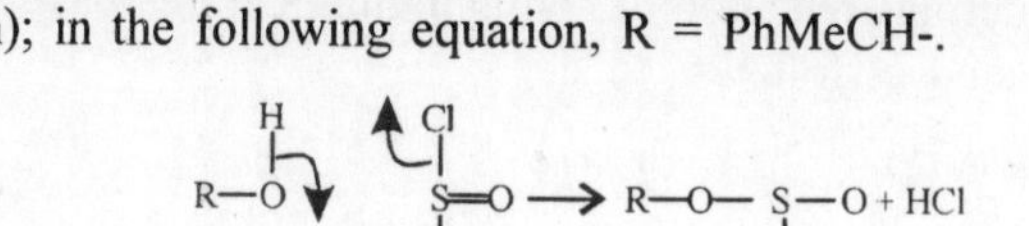

$$\xrightarrow{\text{slow}} \overset{\delta-}{Cl}\text{---R---}\overset{\delta+}{OSO} \xrightarrow{\text{fast}} Cl-R + SO_2$$

(ii) S_N1. This occurs with inversion and retention (racemisation).

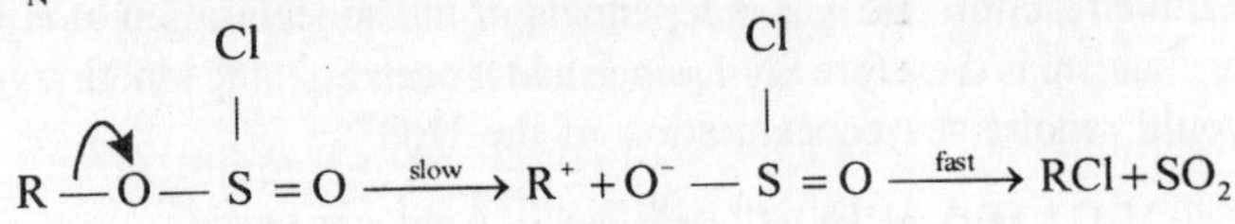

The second stage may possibly be

$$O^- - \underset{}{\overset{Cl}{\overset{|}{S}}} = O \xrightarrow{\text{fast}} SO_2 + Cl^- \xrightarrow[\text{fast}]{R^+} RCl.$$

(iii) S_Ni. This occurs with retention (the reaction is effectively a four-centre type).

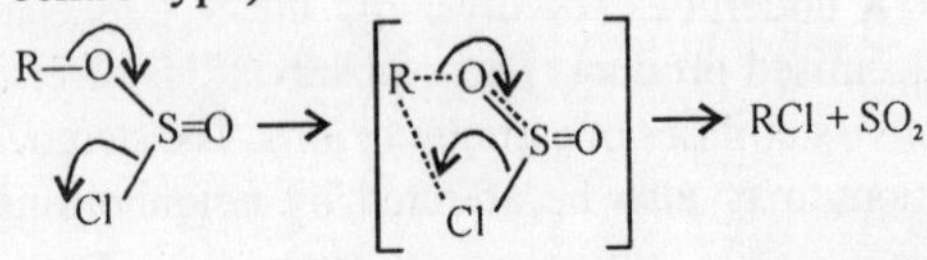

In practice, the α-chloroethylbenzene obtained has almost complete retention of configuration, and consequently the mechanism must be S_Ni.

There appears to be some doubt that the S_Ni mechanism is a one-step process. It has been suggested that the S_Ni mechanism; like the S_N1 mechanism can proceed via ion-pairs, the amount of ionisation depending largely on the nature of the solvent, *e.g.*, the rate of decomposition of chlorosulphites increases with increasing polarity of the solvent. Thus, S_N1 and S_Ni mechanisms may be regarded as extremes, the latter operating when the product has retention of configuration. On this basis, the S_Ni may be formulated as follows:

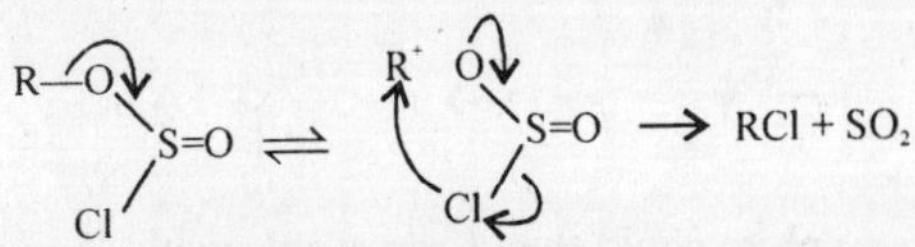

Because of the geometry of the ion-pair, the chlorine atom is forced to attack the carbonium ion from the sane side as the original R—O bond, with consequent retention of configuration.

The formation of the carbonium ion (ion-pair or otherwise) in the S_N1 mechanism is supported by much experimental work, *e.g.*, the chlorosulphite of 3-methylbutan-2-ol, on heating, gives t-amyl chloride

(Lee *et al.*, 1961). This is readily explained by the formation of initially a secondary carbonium ion which then rapidly rearranges to the more stable tertiary carbonium ion.

$$Me_2CH-CHMe-OSOCl-[Me-\underset{\displaystyle Me}{\underset{|}{CH}}-\overset{+}{C}H-Me]\overset{+}{O}SOCl \rightarrow$$

$$[Me-\underset{\displaystyle Me}{\underset{|}{\overset{+}{C}}}-CH_2Me]\overset{-}{O}SOCl \rightarrow Me_2CClCH_2Me+SO_2$$

When α-phenylethanol and thionyl chloride react-in the presence of pyridine, the α-chloroethylbenzene obtained has the inverted configuration (Hughes, Ingold *et al.*, 1937). The explanation offered is that the S_N2 mechanism is operating, the substrate now being a pyridine complex:

$$ROSOCl + C_5H_5N \rightarrow Cl^- + ROSO\overset{+}{N}C_5H_5$$

$$\rightarrow \overset{\delta-}{Cl}---R---\overset{\delta-}{O}SO\overset{+}{N}C_5H_5 \rightarrow Cl—R + SO_2 + C_5H_5N$$

Another example of the S_Ni mechanism (*i.e.*, with retention of configuration) is the decomposition of alkyl chloroformates (Kenyon *et al.*, 1933).

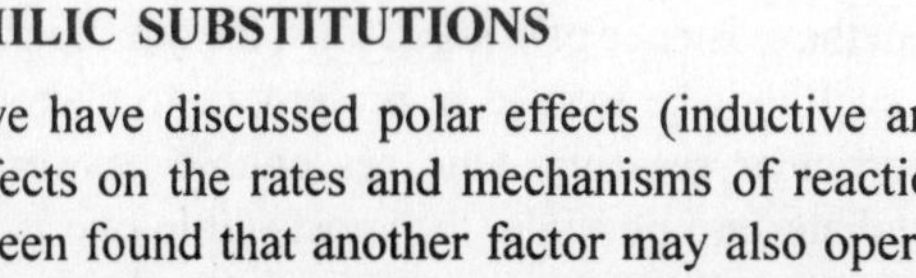

PARTICIPATION OF NEIGHBOURING GROUPS IN NUCLEOPHILIC SUBSTITUTIONS

So far, we have discussed polar effects (inductive and resonance) and steric effects on the rates and mechanisms of reactions. In recent years it has been found that another factor may also operate in various reactions. This factor is known as neighhouring .group participation. Here we have a group attached to the carbon atom adjacent to the carbon atom where nucleophilic substitution occurs and, during the course of the reaction, becomes bonded or partially bonded to the reaction centre to form a non-classical or bridged ion. Thus the rate and/or the stereochemistry of a reaction may be affected. When a reaction is accelerated by neighbouring group participation, that reaction is said to be anchimerically assisted (Winstein *et al.*, 1953). For anchimeric assistance to occur, the neighbouring group, which behaves as a nucleophilic reagent, must be suitably placed stereochemically with respect to the group that is ejected; this is the trans-configuration. Neighbouring group participation is also of great importance in the 1,2-shifts. As we shall see below, neighbouring group participation may also involve a group further removed than the carbon atom adjacent to the reactive centre.

In order to measure anchimeric assistance (kinetic acceleration), it is necessary to be able to estimate reaction rates that would be obtained had there been no neighbouring group participation. This can frequently be done with cyclic structures, since the geometry of the molecule often restricts the possibility of neighbouring group participation to the trans-isomer, the unassisted rate therefore corresponding to that for the cis-isomer. Acyclic structures are much more difficult to study because it is not easy to separate anchimeric assistance from polar and steric effects.

Because of the difficulties in estimating reaction rates in the absence of neighbouring group participation, it is necessary to have a reasonably large increase in rate to justify the conclusion that the rate increase is due to anchimeric assistance only. It appears that a five-fold increase has been accepted as a minimum, but obviously the case is stronger if much larger values are obtained.

In addition to anchimeric assistance as evidence for the formation of non-classical carbonium ions, other criteria have' been used, *e.g.*, stereochemistry of the reaction and structure of the products. In the latter case, the product differs from that which would be expected in the absence of neighbouring group participation.

If we accept the existence of bridged ions, the question to be answered is why should such ions be formed in preference to classical carbonium ions in any particular reaction. One reasonable answer is that when several intermediates are possible, the most stable one is the one likely to be formed. Since charge is more diffuse in the bridged ion than the classical ion, the former would be expected to be more stable than the latter. Whether a group Z can enter into neighbouring group participation depends on the nature of Z, and for a given Z it usually depends on the size of the ring that can be formed (by n.g.p.). The following sections illustrate some examples of neighbouring group participation.

Neighbouring Carboxylate Anion

Hughes, Ingold *et al.* (1937) studied the following reaction of methyl D-a-bromopropionate

$$McCHBrCO_2Me \rightarrow McCH(OMe)CO_2Me$$

With concentrated methanolic sodium methoxide, the reaction was shown to be S_N2, and the product was L-methoxy ester (100 per cent inversion). Under these conditions, the nucleophilic reagent is the methoxide ion, and the reaction is first order with respect to both methoxide ion and ester. When the ester was subjected to methanolysis, *i.e.*, methanol was the solvent (no methoxide ion now present), the product was again L-methoxy ester (100 per cent inversion).

The reaction was now first order (*i.e.*, pseudo first order), but still S_N2, the nucleophilic reagent being the solvent molecules of methanol. When the sodium salt of D-α-bromopropionic acid was hydrolysed in dilute sodium hydroxide solution, the mechanism was shown to be S_N1, and the product was now D-α-hydroxypropionate anion (100 per cent retention). In concentrated sodium hydroxide solution, however, the mechanism was S_N2 (due to the high concentration of the hydroxide ion), and the product was L-α-hydroxypropionate anion (100 per cent inversion).

The explanation for retention is uncertain, but a favoured theory is that an a-lactone is formed first (Kenyon *et al.*, 1936), with expulsion of the bromide ion. Thus, the side opposite to that of the expelled bromide ion is protected from attack by the hydroxide ion, which is consequently forced to attack from the same side as that of the expelled bromide ion, thereby leading to retention of configuration.

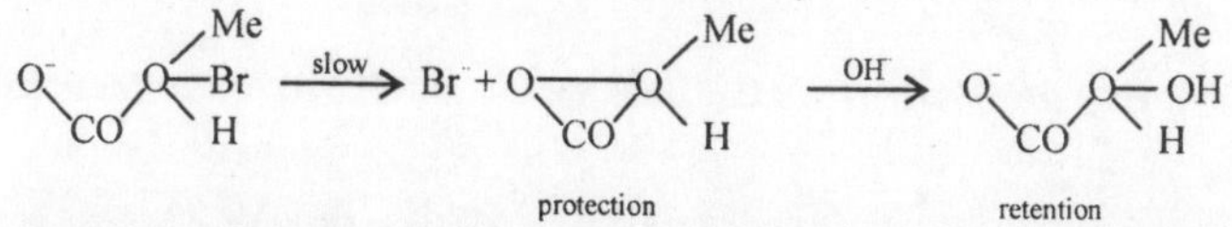

Hughes, Ingold *et al.* (1950) showed that the deamination of optically active alanine by nitrous acid gave an optically active lactic acid with retention of configuration. This is also explained by neighbouring group participation of the a-carboxylate anion:

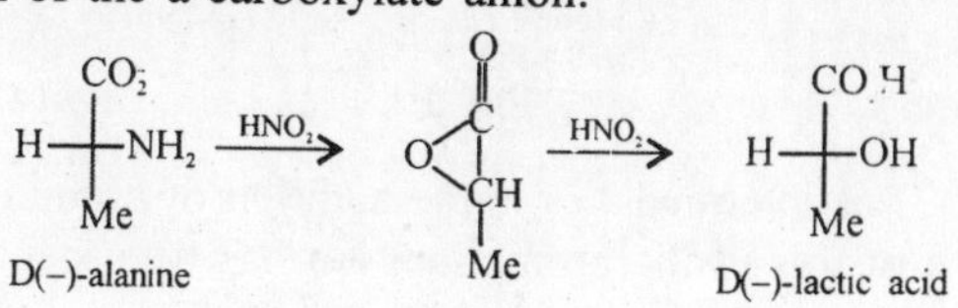

A good example of the effect of ring-size in neighbouring group participation involving the carboxylate ion is shown by the anions of the bromocarboxylic acids:

(I) 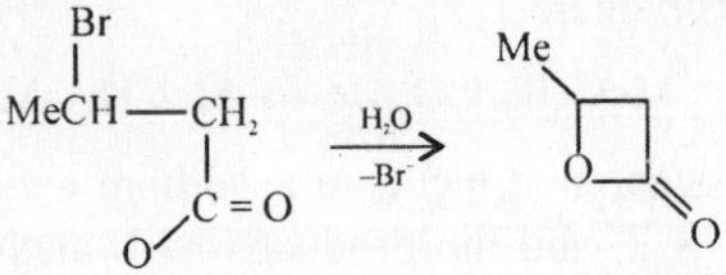

(II) 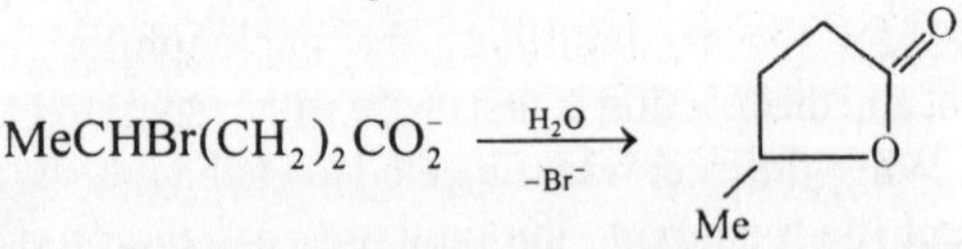

(III) $$Br(CH_2)_6CO_2^- \xrightarrow[-Br^-]{H_2O} HO(CH_2)_6CO_2^-$$

Lactones containing a four- to seven-membered ring can be formed under these experimental conditions, but not eight-membered rings (and larger ones). Whether a three-membered α-lactone ring is actually formed for α-bromopropionic acid is still uncertain, but such a lactone has been isolated for 2-butyl-2-bromohexanoic acid. In this case, presumably steric effects of the large butyl groups prevent attack of the hydroxide ion. At the same time, these large groups will tend to stabilise the small ring (through 'squeezing' the angle O–CHBu$_2$–CO).

Neighbouring Halogen Atoms

Brominium (bromonium) ions were first proposed by Roberts and

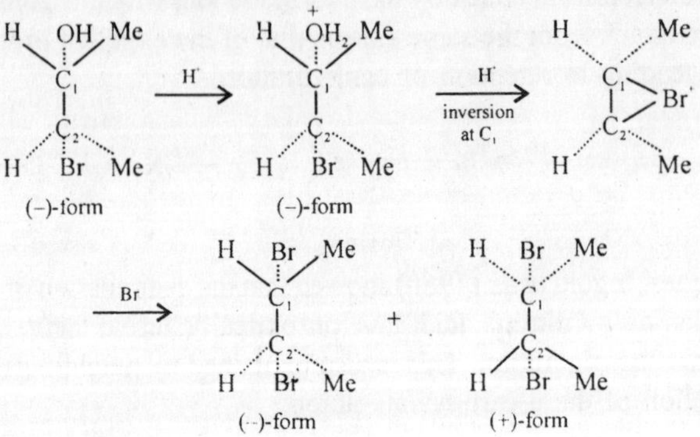

Fig. 1

Kimball (1937) as intermediates in the addition of bromine to alkenes. The existence of this cyclic brominium ion has been demonstrated by

Winstein and Lucas (1939), who found that the action of fuming hydrobromic acid on (–)-threo-3-bromobutan-2-ol gave (±)-2,3-dibromobutane. If no neighbouring group participation of bromine occurred in the above reactions, then if the reaction were S_N2; complete inversion would have occurred only at C_1. If the reaction were the ordinary S_N1, the C_1 would have been a classical carbonium ion (flat), and so inversion and retention (racemisation) would have occurred only at C_1. Since either retention or inversion occurs at both C_1 and C_2, the results are explained by neighbouring group participation of the bromine atom.

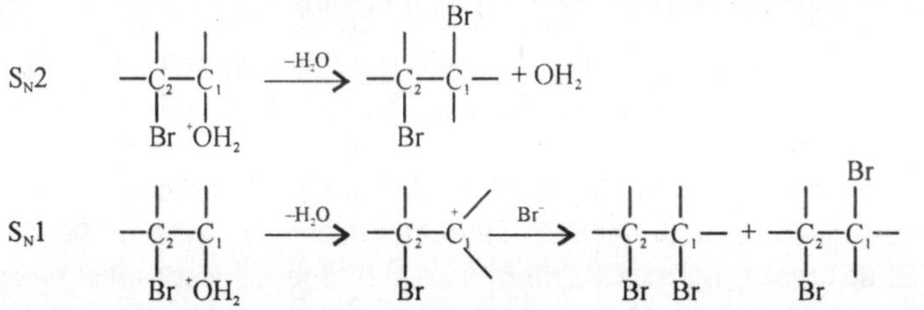

Fig. 2

The above mechanism also explains the formation of meso-2,3-dibromobutane by the action of fuming hydrobromic acid on optically active erythro-3-bromobutan-2-ol (I); (II) and (111) are identical and correspond to the meso-form.

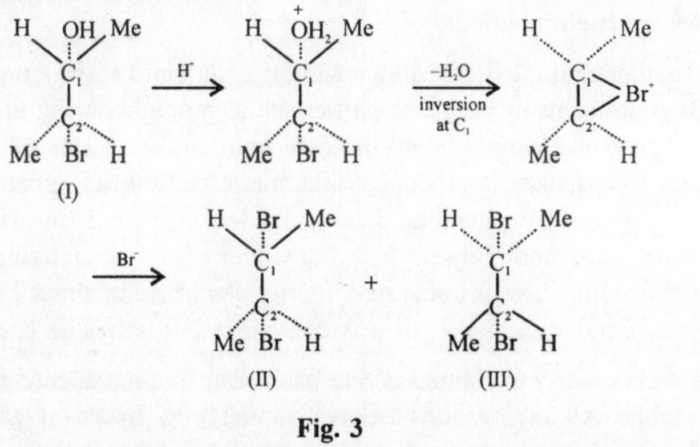

Fig. 3

There is evidence that all the halogen atoms can form cyclic ions and offer anchimeric assistance, *e.g.*, Winstein *et al.* (1948, 1951) studied the acetolysis of cis- and trans-2-halogeno-cyclohexyl brosylates (*i.e.*, p-bromobenzenesulphonates; this group is often written as OBs):

In the absence of neighbouring group participation, the rates would be expected to be about the same. If participation occurs, then this is readily possible in the trans-isomer (I a,2a) by attack of X at the rear of the ejected OBs^- ion, but this is not so for the cis-isomer. The rate ratios observed were :

trans/cis: X = I, $2.7 \times 10^6/1$; X = Br, 800/1; X = Cl, 3.8/1.

Thus iodine affords the greatest anchimeric assistance and chlorine the least.

Neighbouring Hydroxyl or Alkoxyl Group

Hydroxyl groups may enter into neighbouring group participation, one way being via the alkoxide ion. Thus, Bartlett (1935) showed that alkali converts trans-2-chlorocyclohexanol into cyclohexene oxide, and proposed a mechanism in which an alkoxide ion is formed first and this then ring-closes with ejection of the chloride ion: Bergvist (1948) showed that this reaction proceeds more than 100 times as fast as that when the cis-compound is used. Here again, the trans-form permits ready attack at the rear of the chloride ion whereas the cis-isomer does not.

The fact that the cis-form does react may be explained by assuming that the reaction proceeds via cis-elimination of the chlorine atom. This would require a distorted (*i.e.*, highly strained) transition state, and consequently the activation energy for this path would be greater than that for trans-elimination.

Another example is the conversion of sugars into epoxy-sugars. The hydroxyl group itself may also participate as a neighbouring group, the most important example being the case with chlorohydrins of the type $Cl(CH_2)_nOH$. In these compounds, anchimeric assistance is greatest when n = 4 (to give a five-membered ring), far less for n = 5 (to give a six-membered ring), and is absent for other values of n. Thus, tetramethylene chlorohydrin in water is converted into tetrahydrofuran about 103 times as fast as ethylene chlorohydrin is converted into ethylene epoxide.

This is readily explained on the basis' that five-membered rings are very stable. Alkoxyl groups behave similarly to hydroxyl groups in neighbouring group participation, *i.e.*, show anchimeric assistance in the formation of five-membered rings, and to a far less extent in the formation of six-membered rings. Thus, the acetolysis of 4-methoxybutyl brosylate is about 650 times as fast as that of n-butyl brosylate. It should be noted, however, that there appears to be some doubt about this explanation.

Neighbouring Acetoxyl Group

Winstein et at. (1942, 1943) showed that a neighbouring acetoxyl group leads to the formation of an acetoxonium ion. trans-2-Acetoxycyclohexyl brosylate (I) forms trans- l,2-diacetoxycyclohexane (II) when treated with silver acetate, and the same product (II) is obtained when the starting material is trans-2-acetoxycyclohexyl bromide (III). The authors believe that the course of the reaction, based on the stereochemical evidence, proceeds through the same acetoxonium ion (IV).

This mechanism is supported by the fact that in each case, when the reaction was carried out in the presence of a small amount of water, the product was now the monoacetate of cis-cyclohexane-l,2-diol (V); some diacetate of this cis-diol was also obtained. Further support for the formation of (IV) is afforded by the fact that the cis-isomers of (1) and (III) undergo the same reactions but at much slower rates. The formation of the intermediate (Va) is supported by the fact that when the solvolysis of (I) is carried out in ethanol, (VI) is obtained (Winstein *et al.*, 1943).

Neighbouring Phenyl Group

One type of molecular rearrangement is the 1,2-shift, and this may involve neighbouring group participation. An example is the acetolysis of the tosylates (p-toluenesulphonates) of 3-phenylbutan-2-ol (Cram, 1949). If the mechanism were S_N2, then each tosylate would produce an inverted active acetate, inversion occurring only at the C–OTs carbon atom, the result being that the threo-tosylate would give the erythro-acetate and the erythrotosylate the threo-acetate. If the mechanism were SN1, ionisation of the tosyloxy group would produce the classical carbonium ion and the expected result would be racemisation (and retention) at the C–OAc carbon atom, the net result being that both threo- and erythro-tosylate yield the same product.

Cram found that the L-threo-tosylate gave the racemic threo-acetate, whereas the L-ervthro tosylate produced the almost optically pure L-erythro-acetate. Hence the mechanism cannot be S_N2 or S_N1. These results, however, can readily be explained on the basis of the formation of an intermediate bridged carbonium ion, the phenonium cation. This phenonium cation is symmetrical, and since attack by an acetic acid molecule can be expected to occur with e ̣ual probability at carbons 1 and 2, products (1) and (2) will be formed in equal amounts. As these are mirror images, the result is therefore the racemic modification. The phenonium cation from the L-erythro-tosylate is also attacked at carbons

1 and 2 with equal probability, and gives the sane optically active product [(1) = (2)] which has the same configuration as the starting material. The above explanation is completely satisfactory for the results given, but is no longer so when other experimental work is also considered. In acetic acid, the L-threo-tosylate racemises at a rate which is considerably faster than the rate of formation of p-toluenesulphonic acid, and it was estimated that 80 per cent of the tosylate is racemised before its conversion into the acetate (Winstein *et al.*, 1952).

It was therefore suggested that an ion-pair is formed, and this can undergo racemisation through internal return prior to the formation of the postulated phenonium ion which leads to the products. Not only are acetates formed in this reaction, but so are alkenes (25-35 per cent), and when the structures of these alkenes are considered, the results are not explained on the basis of a phenonium ion intermediate. Cram (1952) therefore proposed that a classical cation is formed as well as the bridged ion. The mechanism was then further elaborated to explain the formation of 4 per cent of erti thro-acetate from threo-tosylate. (This formation of erythro- from the threo-compound has been termed 'leakage' from the threo to the erythro system).

Brown *et al.* (1965) have examined this reaction and have provided evidence to show that the results can be explained in terms of a rapidly equilibrating pair of classical ions: In this case, the phenonium ion may be regarded as the transition state formed from either classical carbonium ion and not as the intermediate proposed above.

Another example involving the phenyl group is the ethanolysis of the conjugate base of 2-(p-hydroxyphenyl)ethyl bromide (I), which is about 106 times as fast as that of the corresponding p-methoxy-compound (II) (Winstein *et al.*, 1963). The oxygen atom in (I) is much more electron-releasing than that in (II), and so the intermediate (Ia, which has been isolated) is much more stable than (IIa). Hence, reaction involving the formation of (Ia) requires a lower energy of activation than that involving the formation of (IIa), and so the former proceeds faster than the latter. This, of course, is based on the argument that the formation of (Ia) or (IIa) is the rate-determining step.

Some Other Examples of Neighbouring Group Participation

Many nucleophilic groups can enter into neighbouring group participation but, as we have seen, the extent depends on the structure of the molecule. The sulphur compound (I) undergoes hydrolysis in aqueous dioxan about 10^4 times

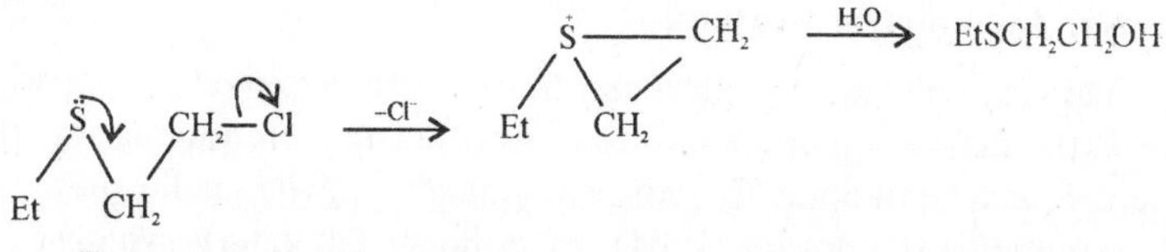

(I)

as fast as that of the oxygen analogue. As we have seen, alkoxyl groups exhibit neighbouring group participation only when the formation of a five- (and six-) membered ring is involved.

Similarly, the alkaline hydrolysis of (III) proceeds much faster than that of n-butyl bromide (to n-butanol); the five-membered ring is stable. On the other hand, the alkaline hydrolysis of (IV) produces the rearranged product (V). The course of this reaction is readily explained on the basis that a cyclic intermediate is formed and that this

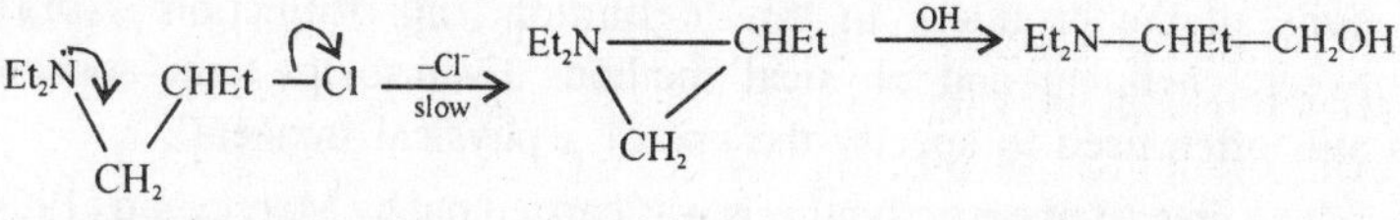

(IV) (V)

undergoes attack at the CH_2 group rather than at the CHEt group because of less steric effects at the former carbon atom.

Double bonds can also enter into neighbouring group participation, *e.g.*, the acetolysis of 4-methylpent-3-enyl tosylate proceeds about 1 200 times as fast as that of ethyl tolylate. The products are 2-cyclopropylpropene and 4-methylpent-3-enyl acetate

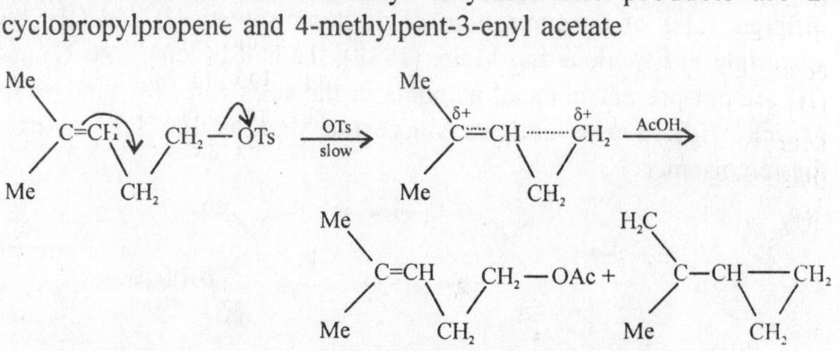

Fig. 4

The intermediate ion (VI) is the homoallylic cation

ASYMMETRIC SYTHESIS (ASYMMETRIC INDUCTION)

Partial Asymmetric Synthesis

Partial asymmetric synthesis may be defined as a method for preparing optically active compounds from symmetric compounds by the intermediate use of optically active compounds, but without the necessity of resolution (Marckwald, 1904). In ordinary laboratory syntheses, a symmetric compound always produces the racemic modification.

This definition included the term 'partial' in order to distinguish this type of asymmetric synthesis from another type in which a `physical reagent' was used instead of optically active compounds. This physical reagent was circularly polarised light, and in this case the process was labelled an 'absolute asymmetric synthesis'. It is now common practice to regard asymmetric synthesis as a special case of stereoselectivity in which a prochiral unit (centre or face) is converted into a chiral centre and results in unequal amounts of stereoisomers (with concomitant optical activity of the product). In this 'definition', no distinction is drawn between 'chemical' and 'physical' methods. Even so, the term 'absolute' is still often used to specify the use of a physical 'reagent'.

The first asymmetric synthesis was carried out by Marckwald (1904), who prepared an active (–)-valeric acid (laevorotatory to the extent of about 10 per cent of the pure compound) by heating the half-brucine salt of ethylmethylmalonic acid at 170°C.

(I) and (II) are diastereoisomers; so are (III) and (IV). (V) and (VI) are enantiomers, and since the mixture is optically active, they must be present in unequal amounts. Marckwald believed this was due to the different rates of decomposition of diastereoisomers (I) and (II), but according to Eisenlohr and Meier (1938), the half-brucine salts (I) and (II) are not present in equal amounts in the solid form (as thought by Marckwald). These authors suggested that as the less soluble diastereoisomer

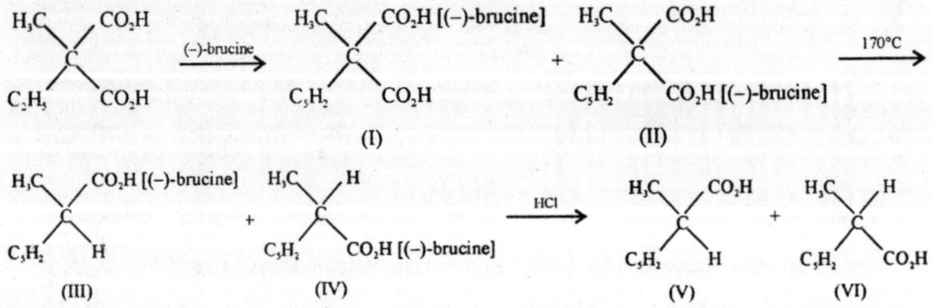

crystallised out (during evaporation of the solution), some of the more soluble diastereoisomer spontaneously changed into the less soluble diastereoisomer to restore the equilibrium between the two; thus the final result was a mixture of the half-brucine salt containing a larger proportion of the less soluble diastereoisomer. If this be the explanation, then we are dealing with an example of asymmetric transformation and not of asymmetric synthesis. Further work, however, has shown that Marckwald had indeed carried out an asymmetric synthesis. Kenyon and Ross (1951) decarboxylated optically active ethyl hydrogen ethylmethyl-malonate (VII) and obtained an optically inactive product, ethyl (±)-a-methylbutyrate (VIII).

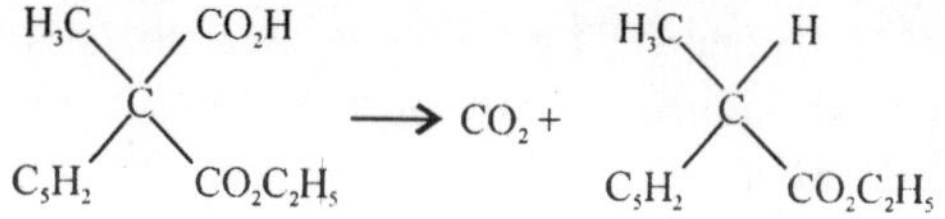

(VII) active (VIII) inactive

These authors (1952) then decarboxylated the cinchonidine salt of (VII) and still obtained the optically inactive product (VIII).

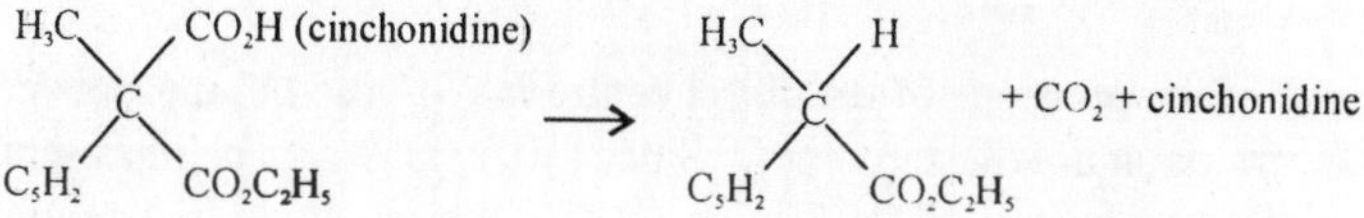

(VIII) inactive

Kenyon and Ross suggested the following explanation to account for their own experiments and for those of Marckwald. Decarboxylation of diastereoisomers (I) and (II) takes place via the formation of the same carbanion (Ia), and decarboxylation of (VII) and its cinchonidine salt via (VIIa).

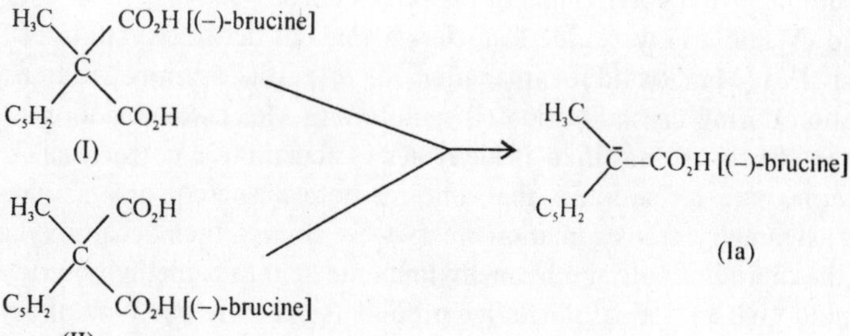

(I) (II) (Ia)

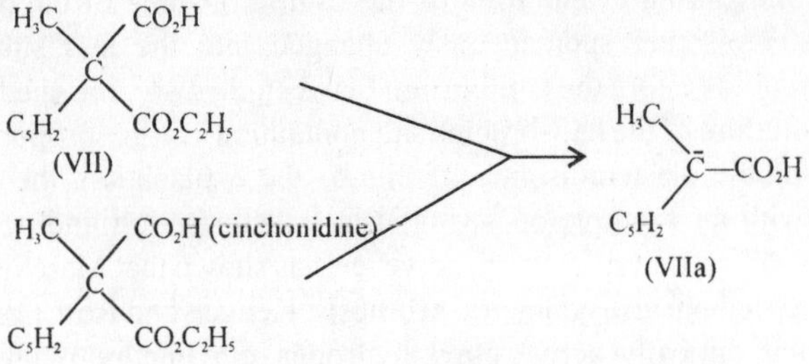

Fig. 5

Combination of carbanion (Ia) with a proton will produce diastereoisomers (III) and (IV) in different amounts, since, in general, diastereoisomers are formed at different rates. Since the carbanion may be represented as a resonance hybrid of (Ia) and (Ib), this carbanion is essentially flat. Also,

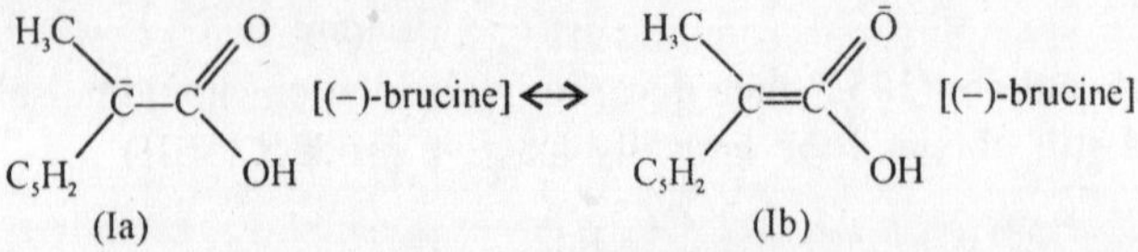

because of the presence of the chiral centre in (–)-brucine, the two faces of the carbanion are diastereotopic. Since (VIIa) may also be represented as a resonance hybrid of (VIIa) and (VIIb), the molecule is essentially flat and has enantiotopic faces.

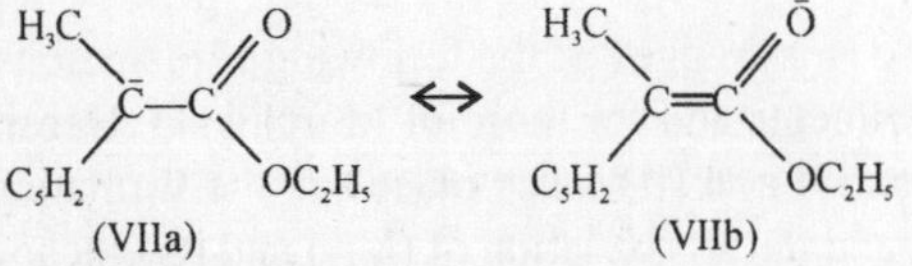

Hence, carbanion (VIIa) will give equimolecular amounts of the enantiomers of (VIII). If the formation of optically active α-methylbutyric acid (V and VI) were due to different rates of decarboxylation of (III) and (IV) (Marckwald's explanation) or to partial asymmetric transformation during crystallisation (Eisenlohr and Meier's explanation), then these effects are nullified if Kenyon's explanation is correct, since the intermediate carbanion is the same for both diastereoisomers. Thus, if the asymmetric transformation theory were correct, then decarboxylation of the dibrucine salt of ethylmethylmalonic acid to α-methylbutyric acid should give an optically inactive product, since only one type of crystal is now possible (asymmetric transformation is now impossible).

(Ia)

On the other hand, if the carbanion (Ia) is an intermediate in this decomposition, it is still possible to obtain an optically active product. Kenyon and Ross did, in fact, obtain a laevorotatory product.

McKenzie (1904) carried out a number of asymmetric syntheses by reduction of the keto group in various keto-esters in which the ester group contained a chiral group, *e.g.*, benzoylformic acid was esterified with (–)-menthol, the ester reduced with aluminium amalgam, and the resulting product hydrolysed; the mandelic acid so obtained was slightly laevorotatory.

$$C_6H_5COCO_2H + (-)\text{-}C_{10}H_{19}OH \rightarrow C_6H_5COCO_2C_{10}H_{19} + H_2O \xrightarrow{Al/Hg}$$

$$C_6H_5CHOHCO_2C_{10}H_{19} \xrightarrow{H_2O} \underset{(-)\text{-rotation}}{C_6H_5CHOHCO_2C} + (-)\text{-}C_{10}H_{19}OH$$

Similarly, the pyruvates of (–)-menthol, (–)-pentyl alcohol and (–)-borneol gave an optically active lactic acid (slightly laevorotatory) on reduction.

$$CH_3COCO_3R(-) \xrightarrow{[H]} CH_3CHOHCO_2R(-) \xrightarrow{H_2O}$$

$$\underset{(-)\text{-rotation}}{CH_2CHOHCO_2H} + (-)\text{-ROH}$$

McKenzie (1904) also obtained similar results with Grignard reagents, *e.g.*, the (–)-menthyl ester of benzoylformic acid and methylmagnesium iodide gave a slightly laevorotatory atrolactic acid.

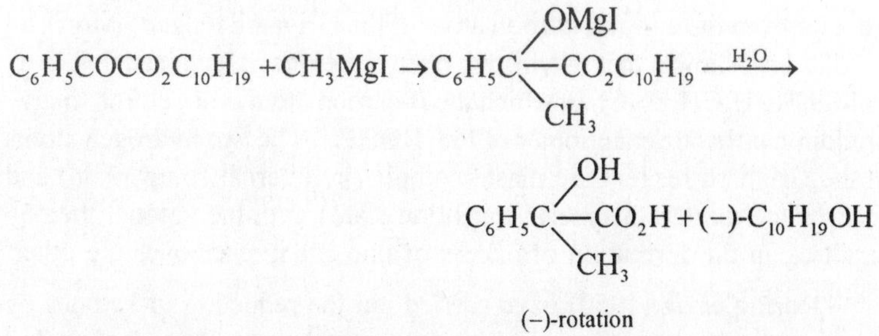

(–)-rotation

Another example of asymmetric synthesis involving the use of a Grignard reagent is the reduction of 3,3-dimethylbutan-2-one into a dextrorotatory 3,3-dimethylbutan-2-ol by means of (+)-2 methylbutylmagnesium chloride (Mosher *et al.*, 1950). The authors explained the result by

$$(CH_3)_3CCOCH_3 \xrightarrow{(+)-CH_3CH_2CH(CH_3)CH_2MgCl} \underset{(+)-\text{rotation}}{(CH_3)_3CCHOHCH_3}$$

postulating the formation ofa cyclic transition state in which the preferred configuration has the methyl group.of the optically active Grignard reagent on the same side of the ring as the larger of the two groups in the ketone. This configuration of the T.S. is preferred because steric

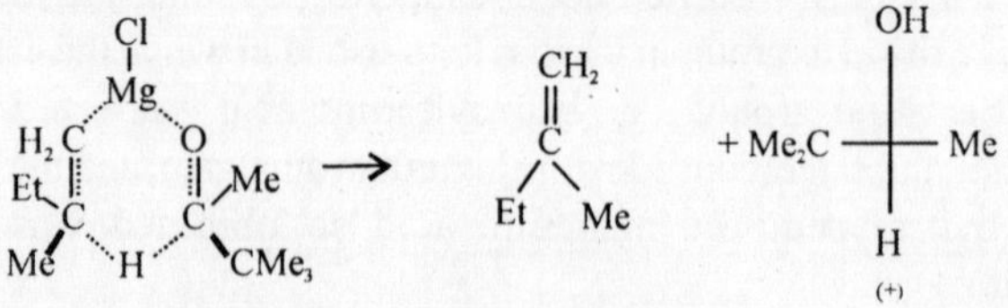

Fig. 6

effects are less than in the alternative T.S., and so the energy content of the former is lower than that of the latter. Mosher used a variety of ketones and obtained a predominance of one enantiomer, but the optical purity was always low (most values were between ~ 4 and 10 per cent). This suggests that the energy differences between the two transition states may be small. Mosher *et al.* (1961, 1966), however, have concluded that the size of the groups is not the only factor that determines the stereo selectivity of these asymmetric, reductions. In fact, although the actual size of a group is fixed, it appears that its effective size depends on the environment and on the mechanism of the reaction. This may lead to difficulties in predicting the steric course of a particular reaction.

In the above examples, the reduction involves a hydrogen atom transfer from the chiral β-carbon atom of the Grignard reagent. Morrison (1967) has shown that isopropyl phenyl ketone is reduced by, *e.g.*, $PhMeCHCH_2CH_2MgCl$, in which the β-carbon atom is not chiral, to give predominantly one enantiomer of $PhCHOHPr^1$. The two hydrogen atoms in the Grignard reagent are diastereotopic (by internal comparison) and so produce diastereoisomeric transition states with the ketone, thereby resulting in the formation of excess of one enantiomer over the other.

Doering *et al.* (1950) have carried out the reduction of ketones by means of the MeerweinPonndorf-Verley method. The reduction of methyl isohexyl ketone with (+)-butan-2-ol in the presence of aluminium

2-butoxide gave (+)-methylisohexylmethanol. Here again the results are explained by the formation of a cyclic transition state with the isohexyl group predominantly on the same side as the methyl group of the butoxide. In both reactions, transfer of a hydride ion takes place.

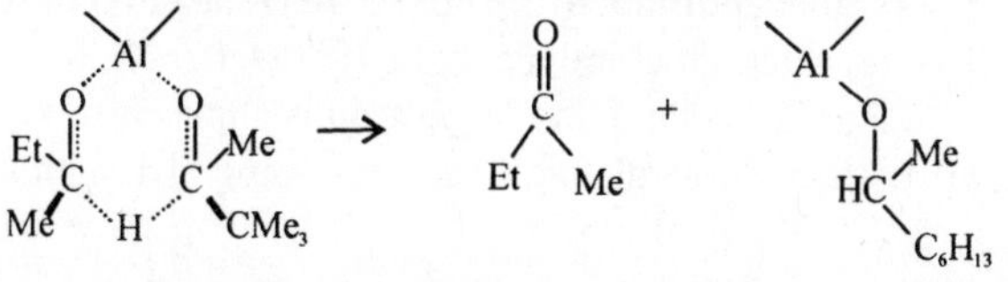

On the other hand, Bothner-By (1951) reduced butanone with lithium aluminium hydride in the presence of (+)-camphor, and thereby obtained (+)-isoborneol (from the camphor) and a small amount of a dextrorotatory butan-2-ol. The reducing agent in this case is a complex aluminohydride ion formed from lithium aluminium hydride and camphor, *e.g.*, $Al(OR)H_3^-$

$$CH_3COC_2H_5 \xrightarrow[(+)\text{-camphor}]{LiAlH_4} \underset{(+)\text{-rotation}}{CH_3CHOHC_2H_3}$$

Turner *et al.* (1949) carried out a Reformatsky reaction (see Vol. I) using acetophenone, (–)menthyl bromoacetate and zinc, and obtained a dextrorotatory β-hydroxy-β-phenylbutyric acid.

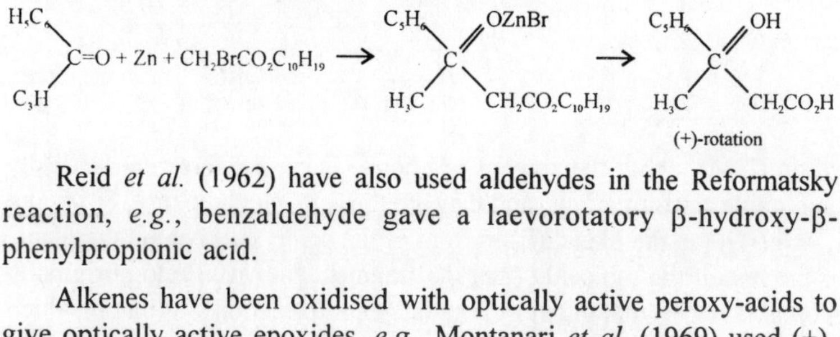

(+)-rotation

Reid *et al.* (1962) have also used aldehydes in the Reformatsky reaction, *e.g.*, benzaldehyde gave a laevorotatory β-hydroxy-β-phenylpropionic acid.

Alkenes have been oxidised with optically active peroxy-acids to give optically active epoxides, *e.g.*, Montanari *et al.* (1969) used (+)-peroxycamphoric acid:

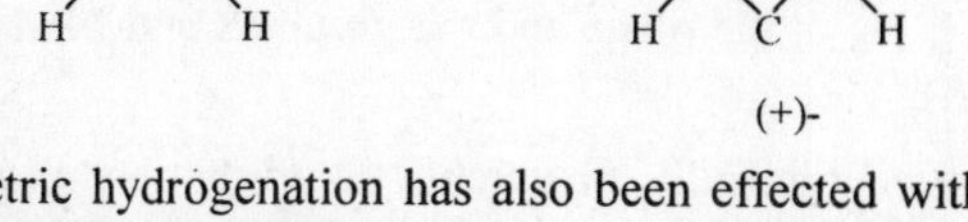

(+)-

Asymmetric hydrogenation has also been effected with the use of a rhodium complex containing, *e.g.*, (+)- or (–)-PhCHMeNHCHO as ligands (McQuillin *et al.*, 1969).

$$PhCMe{=}CHCO2Me \xrightarrow[H_2]{(+)\text{-cat.}} (+)\text{-}PhCHMeCH_2CO_2Me$$

Prelog *et al.* have studied, by means of conformational analysis, the steric course of the addition of Grignard reagents to benzoylformic (phenylglyoxylic) esters of chiral alcohols. If the letters S, M and L refer respectively to small, medium and large groups attached to the carbinol carbon atom of the chiral alcohol, then the general reaction may be written:

$$C_6H_5COCO_2CSML \xrightarrow{RMgX} C_6H_5CR(OH)CO_2CSML$$

$$\xrightarrow{hydrol} C_6H_5CR(OH)CO_2H$$

Prelog *et al.* found that the configuration of the asymmetric carbon atom in the stereoisomer that predominated in this reaction could be correlated with that of the carbinol carbon of the alcohol. The authors considered that the most stable conformation of the ester was the one in which the two oxo groups are planar and trans to each other. It was also believed that the atoms O–C also lie in this plane, *i.e.*, the fragment shown is essentially planar. The problem now was the rotation of the

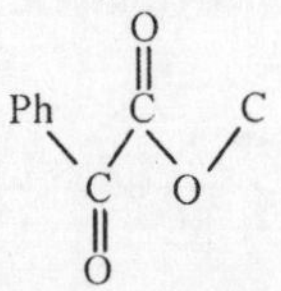

group CSML about the single O–C bond. There are three energetically favourable conformations, and they are those in which, in turn, S, M, and L will (i) lie in the plane; (ii) in front of the plane; (iii) behind the plane, of the rest of the molecule (*i.e.*, the fragment shown). Prelog originally proposed (1953) that the most populated conformation was that in which group S lay in this plane and the groups M and L skew (IXa and Xb; see Chart, in which the thick lines represent groups in front, broken lines groups behind, and ordinary lines groups in the plane). Prelog, however, later (1956) proposed that the most populated conformation was that in which group L lay in the plane and the groups S and M skew (IXb and Xc; see Chart).

The third energetically favourable conformation in which the M group lies in the plane is also shown in the Chart (IXc and Xa). The enantiomeric alcohols (IX) and (X) each give rise to the three energetically favourable conformations for the corresponding ester (see Chart). Another assumption made by Prelog (1953) was that the Grignard reagent attacks the oxo group (of PhCO) from the less hindered side. With methylmagnesium iodide as the Grignard reagent, the resulting x-hydroxy-acid is atrolactic acid (IXd or Xd). The direction of attack is indicated in the

Chart by a thick arrow when attack is from the front face and a broken arrow when attack is from the back. This may be illustrated as shown, with (lXb) as the example.

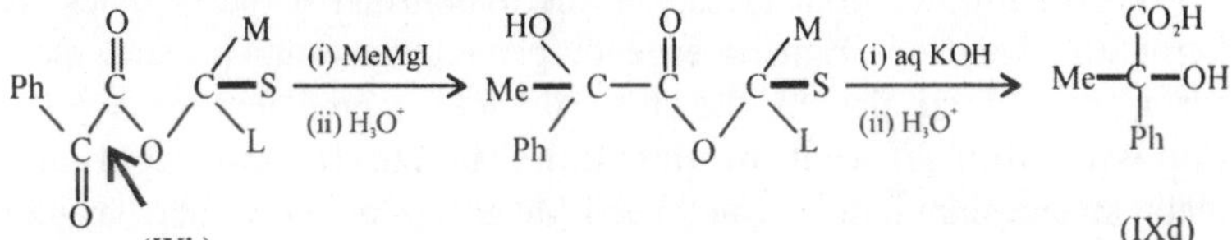

(IXb) (IXd)

Examination of the Chart shows that the predominant enantiomer of atrolactic acid produced is related to the configuration of the chiral alcohol. (IX), via the most populated conformation (IXh), produces (lXd), and (X), via the most populated conformation (Xc), produces (X(l), the enantiomer of (IXd). It might be noted that even if each set of conformations were present in equal populations, the same results are achieved since two of each set of three produce the same atrolactic acid.

Hence, all chiral alcohols of the type CSML(OH) which are stereochemically related, *i.e.*, the groups S, M, and L have the same spatial relationship, will produce the same enantiomer of atrolactic acid in excess. Also, if the sign of rotation of the atrolactic acid produced in excess is measured, the stereochemical relationships of the chiral alcohols can be determined. Now, (–)-menthol, (–)-borneol, and (–)-octan-2-ol are all configurationally related to L(–)-glyceraldehyde and all give (–)-atro lactic acid as the predominant enantiomer. The absolute configurations of these alcohols have been established; they are as shown.

All have the same configuration for groups S, M, and L, and all are alcohols of the type (IX), *e.g.*,

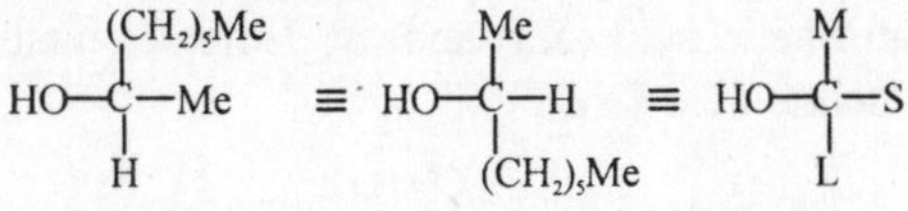

(–)-octan-2-ol (IX)

It therefore follows that (–)-atrolactic acid is (IXd). Hence, if the configuration of the chiral alcohol is known, it is possible to deduce that of the enantiomer of atrolactic produced in excess, and vice versa. This method of correlating configurations has been called the 'atrolactic acid method'. Furthermore, since the absolute configuration of (–)-atrolactic acid is that of (IXd) [determined as described above, and confirmed by other methods], (–)-atrolactic acid is the (R)(–)- and (+)-atrolactic acid is the (S)(+)-acid. Hence, the absolute configurations of chiral alcohols can be correlated with the sign of rotation of the atrolactic acid produced in excess.

It also follows from the above discussion that if the keto-acid is pyruvic acid and the Grignard reagent is phenylmagnesium bromide then, for a given chiral alcohol, the atrolactic acid is predicted to have the opposite sign of that produced by benzoylformic acid and methylmagnesium iodide. The Ph and Me groups are now interchanged; this is illustrated as shown.

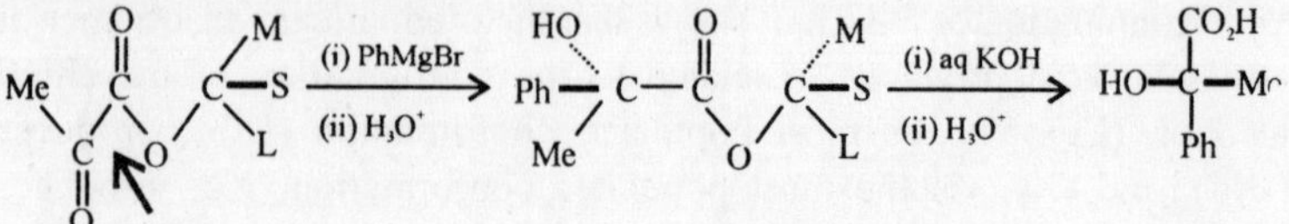

(XI) (Xd)

(XI) corresponds to (IXb), but now atrolactic acid (Xd) is obtained in excess. These predictions have been verified by Prelog *et al.*

An alternative method for the determination of the absolute configuration of a chiral alcohol is that due to Horeau (1961-1964). The alcohol is acetylated in pyridine with an excess of (±)-phenylethylacetic anhydride, $(PhCHEtCO)_2O$, then the excess of the anhydride is hydrolysed and the optical activity of the phenylethylacetic acid is measured. Since one enantiomer of the anhydride reacts with the alcohol faster than the other, the other enantiomer will be in excess in the recovered acid (cf. the kinetic method of resolution.

From the rotation of this acid, the configuration of the alcohol may be deduced from the following empirical rule: The alcohol with the absolute configuration (X) gives, by application of Prelog's method, excess of (+)-atrolactic acid (Xd) and, by Horeau's method, excess of (–)-phenylethylacetic acid (Xe).

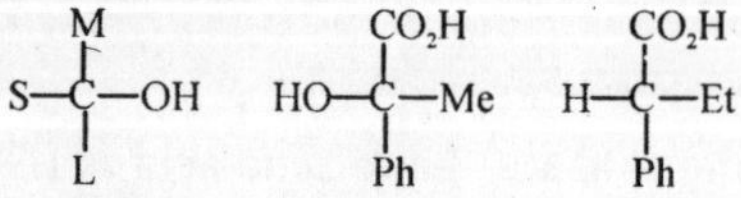

(X) (Xd) (Xe)

Horeau's method is more convenient to carry out than Prelog's method, and has the added advantage in being highly stereoselective.

Cram *et al.* (1952) have also dealt with asymmetric syntheses in which the molecule contains a chiral centre that belongs to the molecule, *i.e.*, remains in the molecule, *e.g.*, the Kiliani reaction in

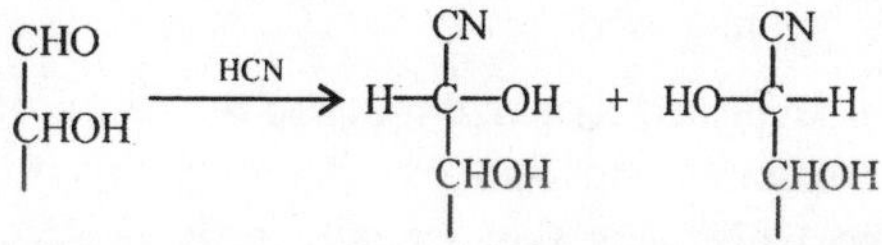

which the two diastereoisomers are formed in unequal amounts. These workers studied in great detail the conversion of acyclic compounds which contained one chiral centre adjacent to an oxo group to the corresponding alcohols, and as a result of their work have formulated the rule of 'steric control of asymmetric induction'.

This is: 'in non-catalytic reactions of the type shown, that diastercoisomer will predominate which would be formed by the approach of the entering group from the least hindered side of the double bond when the rotational conformation of the C–C bond is such that the double bond is flanked by the two least bulky groups attached to the adjacent asymmetric centre.' Thus, using the Newman projection formulae (Z = MgX):

According to Cram's rule, the product from reaction (a) should predominate. An example that demonstrates this is the reaction between α-phenylpropionaldehyde (S = H, M = Me, L = Ph) and methylmagnesium bromide (R^1 = Me); two products can be formed, viz., (XII) (the ervthrocompound) and (XIII) (the threo-compound):

The ervthro-compound (XII) is the major product. Some exceptions to the predictions have been observed, particularly when group L is very large. Other models have been proposed.

Karabatsos (1967), using certain assumptions, has been able to predict semiquantitatively from the Cram model (non-catalytic type) the stereospecificity of the product. As pointed out above, the rule will not apply to catalytic reduction, and it also does not apply to compounds in which the asymmetric carbon atom is joined to a group capable of complexing with the reagent, *e.g.*, OH, NH_2.

The influence of enzymes on the steric course of reactions has also been investigated, *e.g.*, Rosenthaler (1908) found that emulsin converted benzaldehyde and hydrogen cyanide into dextrorotatory mandelonitrile which was almost optically pure. It has been found that in most enzymic reactions the product is almost 100 per cent of one or other enantiomer. Enzymes are proteins and are optically active, but since they are so 'one-sided' in their action, it appears likely that the mechanism of the reactions in which they are involved differs from that of asymmetric syntheses where enzymes are not used.

It has been suggested that enzymes are the cause of the formation of optically active compounds in plants. Although this is largely true, the real problem is: How were the optically active enzymes themselves produced? Ferreira's work, however, shows that optically active compounds may possibly be produced in living matter by activation of a racemic modification. This theory appears to be superior to that of the formation of optically active compounds by the action of naturally polarised light.

Asymmetric syntheses involving elimination reactions have also been carried out to prepare optically active alkenes, *e.g.*, Goldberg *et al.* (1966) subjected (-)-cis-4-methylcyclohexyl hydratropate to pyrolysis and obtained (-)-4-methylcyclohexene (in very low optical purity). Now let us consider reactions in which the products are not chiral but have been formed by two identical groups in the substrate reacting at different rates.

These are highly stereoselective, and since they involve enzymes, may be regarded as a special class of asymmetric synthesis. A classical example of this type is the oxidation of citric acid (XIV) to α-ketoglutaric acid (XV) by means of an enzyme:

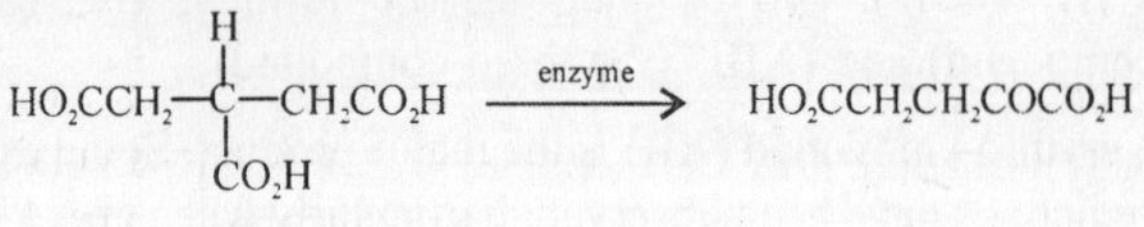

(XIV) (XV)

According to this equation, only the right-hand side methylene group is oxidised; the two groups are enantiotopic and react at different rates with a chiral reagent. This was demonstrated by the use of deuterium-labelled citric acid (XIVa) to give (XVa).

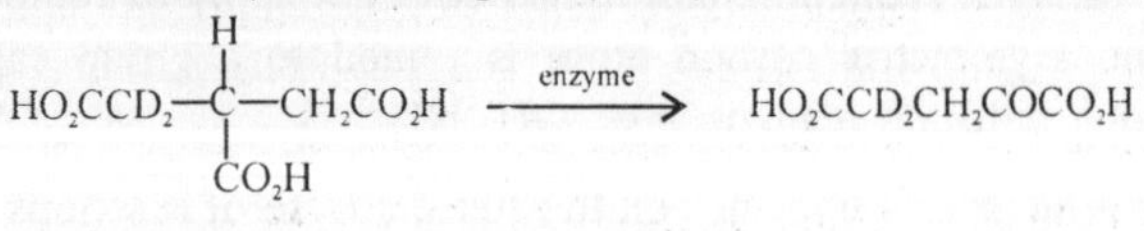

(XIVa) (XVa)

ABSOLUTE ASYMMETRIC SYNTHESIS

As we have seen, right- and left-circularly polarised light is unequally absorbed by enantiomers, provided the light has a wavelength in the neighbourhood of the characteristic absorption bands of the compound.

It has been suggested that circularly polarised light produced the first natural active compounds, and to support this theory, racemic modifications have been irradiated with circularly polarised light and attempts made to isolate one enantiomer. There was very little success in this direction until W. Kuhn and Braun (1929) claimed to have obtained a small rotation in the case of ethyl a-bromopropionate. The racemic modification of this compound was irradiated with right- and left-circularly polarised light (of wavelength 2 800Å), and the product was found to have a rotation of + or –0.05°, respectively.

Thus we have the possibility of preparing optically active products from inactive substances without the intermediate use of optically active reagents (cf. Ferreira's work). This type of synthesis is known as an absolute asymmetric synthesis; it is also known as an absolute asymmetric decomposition (or destruction), since one enantiomer is decomposed (or destroyed) preferentially. On the other hand, there is another type of absolute asymmetric synthesis in which a chiral physical 'reagent' converts an achiral substrate into a chiral product.

From 1930 onward, more conclusive evidence for absolute asymmetric decompositions has been obtained, *e.g.*, W. Kuhn and Knopf (1930) irradiated (±)-α-azidopropionic dimethylamide, $CH_3CHN_3CON(CH_3)_2$, with right-circularly polarised light and obtained an undecomposed product with a rotation of +0.78°; with left-circularly polarised light, the undecomposed product had a rotation of –1.04°. Thus the (–)- or (+)-form is decomposed (photochemically) by right- or left-circularly polarised light, respectively. Similarly, Mitchell (1930) irradiated humulene nitrosite with right- and left-circularly polarised red light, and obtained slightly optically active products.

Davis and Heggie (1935) found that the addition of bromine to 2,4,6-trinitrostilbene in a beam of right-circularly polarised light gave a dextrorotatory product; this is an example of an absolute asymmetric synthesis. Small (+)-rotations were also observed when a mixture of ethyl fumarate and anhydrous hydrogen peroxide in ethereal solution was irradiated with right-circularly polarised light (Davis *et al.*, 1945). As mentioned above, there has been much speculation about how optically active compounds are produced in nature. A difficulty of the 'circularly polarised light' theory is that, up to the present, no satisfactory answer has been offered to explain the source of such light in nature.

On the other hand, Lee *et al.* (1956) showed that electrons emitted in β-decay are polarised, and if they slow down they lose some of their energy by emission of γ-radiation which is circularly polarised. Garay (1968) has irradiated D- and L-tyrosine with β-particles derived from strontium-90 Y and showed that the D-enantiomer was decomposed more quickly. He failed to 'activate' racemic DL-tyrosine, but suggested that β-particle bombardment may he the source of optically active compounds in nature.

ASYMMETRIC SYNTHESIS

The direct (without resoulution) synthesis of an optically ative compounds and most extensively studied type of heterolytic substitution in saturated compounds is the nucleophilic type, *i.e.*, the S_N1 and S_N2 mechanisms.

One-stage Process: When two molecules simultaneously undergo covalency change in the rate-determining step, the mechanism is called bimolecular and is labelled S_N2 (substitution, nucleophilic, bimolecular).

Two-stage Process: In this case the first step is the slow heterolysis of the compound to form a carbonium ion, and this is then followed by the second step of rapid combination of the carbonium ion with the nucleophilic reagent. The rate-determining step is the first, and since in this step only one molecule is undergoing covalency change, the mechanism is called unimolecular and is labelled S_N1 (substitution, nucleophilic, unimolecular).

The symbols S_N1 and S_N2 were introduced by Ingold (1928), the number in the symbol referring to the molecularity of the reaction and not to the kinetic order. Any complex reaction may be designated by the molecularity of its rate-determining stage, the molecularity of the rate-determining stage being defined as the number of molecules necessarily undergoing covalency change (Ingold, 1933). It is also important to note that the definitions of S_N1 and S_N2 mechanisms do not take into account the salvation of the initial molecules and the transition states. Salvation energies play a very important part in determining the activation energies of reactions in solution.

A number of differences exist between S_N2 and S_N1 reactions, *e.g.*:

(i) When both reactants are present in small and controllable concentrations, S_N2 reactions are second-order and S_N1 reactions are first-order. In a bimolecular reaction, if one of the reactants is in constant excess, *e.g.*, one reactant is the solvent, then the mechanism is still bimolecular but the reaction is now of the first order. On the other hand, although the unimolecular mechanism often leads to first-order kinetics, it may, under certain circumstances, follow a complicated kinetic expression.

(ii) The S_N2 mechanism always leads to inversion of the configuration of the products, whereas with the S_N1 mechanism there may be inversion and/or retention, the amount of each depending on various factors.

(iii) No rearrangement is possible with the S_N2 mechanism, but is possible (and often occurs) with the S_N1 mechanism.

(iv) The rate constant of an S_N2 reaction with a given substrate depends on the nature of the nucleophile, and for a given nucleophile, on the nature of the leaving group in the substrate.

The nucleophilic reagent may be negatively charged or neutral; the primary requirement is that it must possess an unshared pair of electrons which it can donate to a nucleus capable of sharing this pair. One widely studied example of nucleophilic aliphatic substitution is that of the hydrolysis of alkyl halides.

$$S_N2 \quad Y^- + R—X \xrightarrow{\text{slow}} \overset{\delta-}{Y}....R....\overset{\delta-}{X} \xrightarrow{\text{fast}} Y—R + X^-$$

$$S_N1 \quad R—X \xrightarrow{\text{slow}} \overset{\delta+}{R}....\overset{\delta+}{X} \xrightarrow{\text{fast}} R^+ + X^-$$

$$R^+ + X^- \xrightarrow{\text{fast}} RY$$

Of particular interest is the evidence for the S_N1 mechanism. A fundamental part of this mechanism is the postulate of carbonium ions as transient intermediates. Triarylmethyl carbonium ions have been isolated as their salts, *e.g.*, triphenylmethyl perchlorate, $Ph_3C^+CiO_4^-$, and borofluoride, $Ph_3C^+BF_4^-$ (Dauben *et al.*, 1960). The stability of ions such as these is attributed to resonance. On the other hand, since the order of stability of alkyl carbonium ions is tertiary > secondary > primary, success is more likely to be achieved in the isolation of tertiary alkyl carbonium ions. Thus, Olah *et al.* (1964) have prepared, *e.g.*, $Me_3C^+SbF_6^-$, and their infrared studies have substantiated the planar sp^2 hybridised structure of the simple alkyl carbonium ions.

A point of interest in connection with the S_N1 mechanism is that it is S_N1 because the ratedetermining step is ionisation of RX. If, however, combination with Y^- is the rate-determining step, then the mechanism is referred to as the $S_N2(C^+)$ mechanism (substitution, nucleophilic, bimolecular, with a rapidly formed carbonium ion; Hughes, Ingold *et al.,* 1954). An example of this type is the Friedel-Crafts reaction involving diarylmethanols and arenes in the presence of a strong acid (Bethell *et al.*, 1958, 1959).

$$Ar_2^1CHOH + H^+ \underset{}{\overset{\text{fast}}{\rightleftharpoons}} Ar_2^1CHOH_2^+ \overset{\text{fast}}{\rightleftharpoons}$$

$$H_2O + Ar_2^1CH^+ \xrightarrow[\text{slow}]{Ar^2H} Ar_2^1CH - Ar^2 + H^2$$

One other point that will be mentioned here is the problem of nomenclature. The term 'carbonium ion' has been commonly used, but other terms have been proposed. Olah (1972) has suggested that the general name 'carbocation' be used for positive ions of carbon compounds (cf. 'carbanions' for negative ions). Carbocations are of two types:

(i) **Carbenium Ions:** These are trivalent ('classical') ions containing an sp2-hybridised electron deficient carbon atom, and tend to be planar.

(ii) Penta- or tetraco-ordinated ('non-classical') ions containing a carbon atom involved in three, two-electron covalent bonds and a fourth two-electron three-centre bond. These are carbonium ions (this is in line with onium ions).

e.g..,

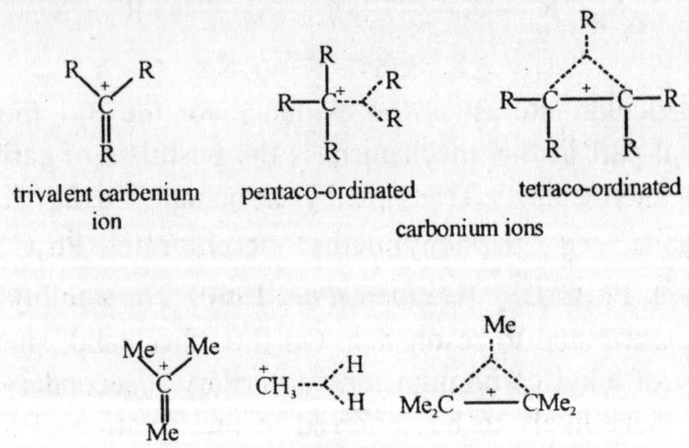

Fig. 7

Carbenium ions have been differentiated from carbonium ions by means of spectroscopic studies, *e.g.*, i.r., NMR (Olah *et al.*, 1970, 1972).

Any factor that affects the energy of activation (E) of a given type of reaction will affect the rate and/or the mechanism. The following discussion is largely qualitative, and because of this, one cannot be sure which are the predominant factors in deciding the energy of activation. We shall discuss, for the hydrolysis of alkyl halides, the influence of the following factors: The nature of R (polar and steric effects); the nature of X and Y; and the nature of the solvent.

The Nature of R

(a) **Polar Effects:** Let us consider the series EtX, i-PrX, and t-BuX. Since the methyl group has a + I effect, the larger the number of methyl groups on the carbon atom of the C-X group, the greater will be the electron density on this carbon atom. This may be represented qualitatively as follows:

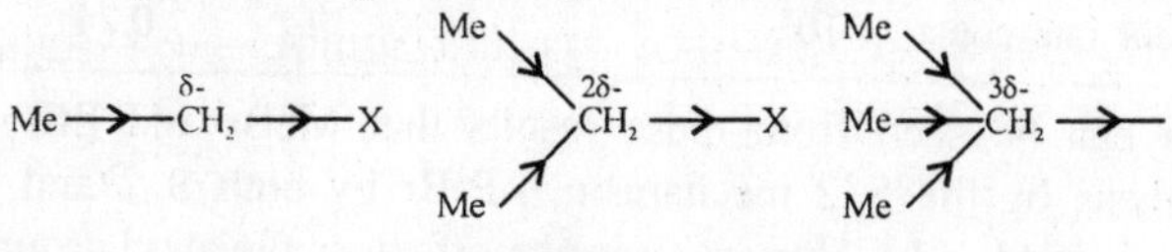

Fig. 8

This increasing negative charge on the central carbon atom increasingly opposes attacks at this carbon by a negatively charged nucleophilic reagent; it also opposes, to a lesser extent, attack by a neutral nucleophilic reagent since this still donates an electron pair. Thus the formation of the transition state for the S_N2 mechanism is opposed more and more as the charge on the central carbon atom increases. (There is also an increasing steric effect operating; this is dealt with in).

The anticipated result, therefore, is that as the number of methyl groups increases on the central carbon atom, the S_N2 mechanism is made more difficult in passing from EtX to t-BuX. On the other hand, since the S_N1 mechanism involves ionisation of RX (in the rate-determining step), any factor that makes easier the ionisation of the molecule will therefore facilitate the S_N1 mechanism. The anticipated result, therefore, is that the greater the negative charge on the central carbon atom, the easier will be the ionisation of RX since X is displaced with its covalent electron pair; thus the tendency for the SN1 mechanism should increase from EtX to t-BuX.

An alternative explanation for the effect of the nature of R is as follows. When the S_N1 mechanism is favoured, it therefore follows the activation energy for this path is less than that for the S_N2; and vice versa. Since the S_N1 reaction proceeds through a carbonium ion, the more stable this ion the lower is the activation energy, and the more favoured will be this mechanism. Now, the order of stabilities of carbonium ions is prim. < s < t, and hence the tendency for the S_N1 mechanism to operate will be t-BuX > i-PrX > EtX, and the reverse tendency for the S_N2 mechanism.

These predicted results have been verified experimentally. Hughes, Ingold *et al.* (1935-1940) examined the rates of hydrolysis of alkyl bromides in alkaline aqueous ethanol at 55°C:

	MeBr	***EtBr***	***i-PrBr***	***t-BuBr***
2nd-order rate const. × 10^5	2140	170	4.7	
1st-order rate const. × 10^5			0.24	1 010

It can be seen from these results that MeBr and EtBr undergo hydrolysis by the S_N2 mechanism, i-PrBr by both S_N2 and S_N1, and t-BuBr by S_N1 only. Thus, as the polar effects in the alkyl group produce an increasing electron density on the central carbon atom, the rate of the S_N2 mechanism decreases and a point is reached where the mechanism changes over to S_N1. With i-PrBr both S_N2 and S_N1 mechanisms operate, and the rate of the S_N2 mechanism is much less than that of the S_N2 mechanism for EtBr. With t-BuBr the electron density on the central carbon atom is so great that the S_N2 mechanism is completely inhibited; a very rapid hydrolysis occurs by the S_N1 mechanism only.

Since the mechanism is S_N1, it therefore means that the hydroxide ion does not enter into the rate-determining step of the hydrolysis. This has been proved as follows. The hydrolysis of t-BuBr was carried out in an alkaline solution containing less than the equivalent amount of hydroxide ion (compared with the alkyl bromide). Thus, although the solution was originally alkaline, as the hydrolysis proceeds, the solution becomes neutral and finally acid; nevertheless, the rate of the hydrolysis was dependent only on the alkyl bromide concentration.

As pointed out above, there are reactions which occur under intermediate conditions, *i.e.*, at the border-line between the extreme S_N1 and S_N2 mechanisms. Some authors believe that in this borderline region there is only one mechanism operating, *e.g.*, Prevost (1958) has postulated, on theoretical grounds, the existence of a more universal 'meso-mechanism'. There is, however, much experimental work in favour of concurrent S_N1 and S_N2 mechanisms operating. Gold (1956) has described evidence for this view, and more recently Swart *et al.* (1961) have shown that the exchange reaction between diphenylmethyl chloride and radiochlorine (as LiCl) in dimethylformamide occurs by a simultaneous S_N1–S_N2 mechanism. On the other hand, Fava *et al.* (1963) studied the following isotopic exchange with methyl cyanide as solvent $(\overset{*}{S} = {}^{35}S)$:

$$Ar_2CHSCN + \overset{*}{S}CN^- \rightarrow Ar_2CH\overset{*}{S}CN + SCN^-$$

The authors used various p- and p, p′-substituted diphenylmethyl (benzhydryl) thiocyanates, and showed that the 4-nitro compound obeyed a second-order rate law and the 4,4′-dimethyl derivative obeyed a first-order rate law. These results were interpreted as indicating. S_N2 and S_N1 mechanisms respectively. The unsubstituted and the 4-chloro compound obeyed mixed first- and second-order rate laws, and the authors interpret this as a region of simultaneous S_N1 and S_N2 mechanisms.

The actual position where the mechanism changes over from S_N2 to' S_N1 in a graded series, *e.g.*, in the alkyl halides, is not fixed but depends on other factors such as the concentration and nature of the nucleophilic reagent, and on the nature of the solvent.

Experimental work has shown that higher n-alkyl groups behave similarly to ethyl. For a given set of conditions, the kinetic order is the same, but the rates tend to decrease as the number of carbon atoms increases, *e.g.*, Hughes, Ingold *et al.* (1946, 1948) showed that the reactions between primary alkyl bromides and ethoxide ion in dry ethanol are all S_N2, and their relative rates (at 55°C) are Me, 17.6; Et, 1.00; n-Pr, 0.31; n-Bu, 0.23; n-pentyl, 0.21.

Similar results were obtained for secondary alkyl groups. In these cases the mechanisms were both S_N2 and S_N1, but the rates for one or other order were reasonably close, *e.g.*, for the second-order reactions of secondary bromides with ethoxide ion in dry ethanol at 25°C, Hughes, Ingold *et al.* (1936-) found that the relative rates were: i-Pr, 1.00; 2-n-Bu, 1.29; 2-n-pentyl, 1.16; 3-n-pentyl, 0.93. These authors also showed that higher tertiary alkyl groups behaved similarly to t-Bu, all showing a strong tendency to react by the S_N1 mechanism.

When hydrogen atoms in methyl chloride are replaced by phenyl groups, the mechanism of the hydrolysis may be changed (from S_N2). The presence of a phenyl group produces a carbonium ion which can be stabilised by resonance; this acts as the driving force to produce ionisation; *e.g.*, Thus, one can anticipate that as the number of phenyl groups increases, the stability of the carbonium ion produced will increase; *i.e.*, the carbonium ion will be formed more readily and consequently the S_N1 mechanism will be increasingly favoured. Thus in the series McCI, $PhCH_2Cl$, Ph_2CHCI, Ph_3CC1, it has been found that in alkaline solution the hydrolysis of methyl chloride proceeds by the S_N2 mechanism, that of phenylmethyl chloride by both S_N2 and S_N1, and that of diphenylmethyl chloride by S_N1; the hydrolysis of triphenvlmethyl chloride is too fast to be measured, but this high rate is very strong evidence for an S_Nl mechanism.

Various groups in the para-position of the phenyl nucleus either assist or oppose ionisation. It has been found that alkyl groups enhance ionisation in the order Me > Et > i-Pr > t-Bu. Since this order is the reverse of that expected from the general inductive effects of these groups, it has been explained by the hyperconjugative effects of these groups. On the other hand, a nitro-group retards the ionisation; and this attributed to the electron-withdrawing effect of this group.

Another 'group' that has a polar effect is deuterium, which appears to be electron- releasing with respect to hydrogen. Thus, deuteration at the a-carbon atom increases the rate of solvolysis in certain types of S_N1 reactions, *e.g.*, the rate of solvolysis of Ph_2CDC1 is about 17 per cent faster than that of Ph_2CHCl. This effect of deuterium is an example of the secondary isotope effect. On the other hand, it appears that substitution of H by D on an a-carbon atom has very little effect on the rates of S_N2 reactions. In some cases, however, the rate is decreased, and it has been suggested that deuteration increases the energy of activation.

Another group of interest is the carbonyl group; this is electron-attracting (through resonance):

$$-\overset{|}{C_1}-\overset{|}{C}=O \leftrightarrow -\overset{\delta+}{\overset{|}{C_1}} \rightarrow \overset{|}{C^+}-O^-$$

Hence, the covalent electron-pair of a halogen atom attached to C_1 is drawn closer to C_1 and consequently it is more difficult for this halogen atom to ionise. Thus the S_N1 mechanism is opposed, and at the same time, the small positive charge on C_1 encourages the S_N2 mechanism. It can therefore be anticipated that any electron-attracting (or withdrawing) group will tend to inhibit the S_N1 mechanism for a compound with an a-halogen atom. Such groups are CO_2R, NO_2, CN, etc.; *e.g.*, both ethyl a-bromopropionate and diethyl bromomalonate undergo hydrolysis by the S_N2 mechanism.

On the other hand, the carboxylate ion has a + I effect due to its negative charge and hence its presence should enhance the ionisation of an a-halogen atom. At the same time, the a-carbon atom tends to acquire a small negative charge, and this will tend to oppose the approach of a hydroxide ion. Thus there are two influences acting, one increasing the tendency for the S_N1 mechanism and the other decreasing the tendency for the S_N2; both therefore oppose the S_N2 mechanism. Some experimental results that illustrate these arguments are the alkaline hydrolyses of the following compounds:

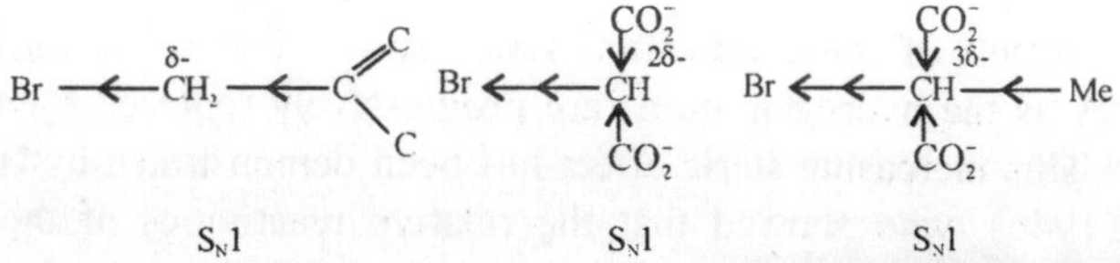

Fig. 9

A point of interest in connection with the S_N1 mechanism is that it is catalysed by heavy metal salts, particularly silver salts. This is believed to be due to the formation of complexes, thereby facilitating ionisation, which is the rate-determining step. Complex-formation occurs by donation of a lone pair of electrons on the halogen atom to an empty orbital of the metal ion.

$$R-\ddot{X} + Ag^+ \underset{\text{fast}}{\overset{\text{fast}}{\rightleftharpoons}} [R-X-Ag]^+ \xrightarrow{\text{slow}} AgX + R^+ \xrightarrow[\text{fast}]{OH} ROH$$

The Nature of R

(b) **Steric Effects:** In the transition state for the SN2 mechanism, there are five atoms or groups bonded or partly bonded to the reaction carbon atom. Thus the larger the bulk of these groups, the greater will be the compression energy (*i.e.*, greater steric strain) in the transition state and consequently the reaction will be sterically hindered. The problem is different for the S_N1 mechanism. Here, the transition state does not contain more than four groups attached to the reaction carbon atom and hence one would expect that steric hindrance should be less important. On the other hand, if the molecule undergoing the S_N1 mechanism contains particularly large groups, then the first step of ionisation may relieve the steric strain and so assist the formation of the carbonium ion, *i.e.*, the reaction may be sterically accelerated.

Let us now examine some examples involving steric effects.

(i) The following series of alkyl halides, MeX, EtX, isoPrX and t-BuX, may be made to undergo the S_N2 mechanism under suitable conditions; the transition state contains three a-bonds (sp^2 hybridisation) in one plane and two partial bonds which are collinear and perpendicular to this plane. Thus we have:

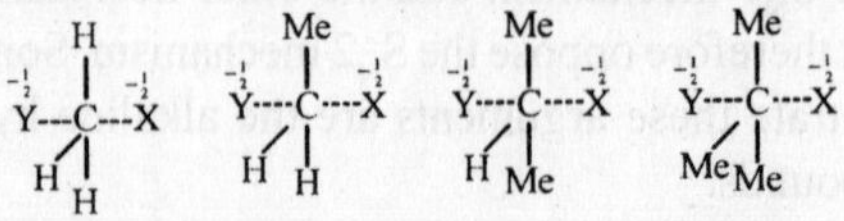

Inspection of these transition states shows that steric hindrance increases as the hydrogen atoms are progressively replaced by methyl groups. This increasing steric effect has been demonstrated by Hughes *et al.* (1946), who showed that the relative reactivities of the alkyl bromides towards iodide ions in acetone (by the S_N2 mechanism) are: Me, 10 000; Et, 65; isoPr, 0.50; t-Bu, 0.039. However, as we saw above the polar effect opposes the S_N2 mechanism in the order t-Bu > isoPr > Et > Me. Hence, this factor will also affect rates in the same direction as steric effects. On the other hand, if we now consider the SN2 mechanism for n-propyl; isobutyl and neopentyl halides, then, since there are no methyl groups on the a-carbon atom, polar effects are almost completely absent (inductive effects fall off very rapidly from the source). Hence any differences in reaction rates among these three compounds may be attributed solely to steric effects.

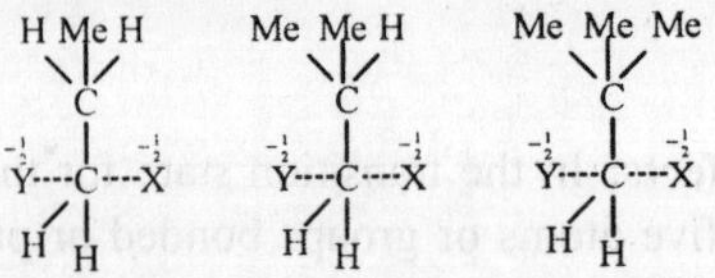

At first sight one would not expect n-PrX to show an added steric effect when compared with EtX since the added methyl group can occupy a position close to the plane of the transition state (*i.e.*, the plane containing the three a-bonds), and so would not offer any appreciable steric hindrance. In practice, however, n-propyl halides are less reactive than the corresponding ethyl halides. Magat *et al.* (1950) have offered the following explanation. The smaller the number of conformations available in the activated as compared with the initial state produces a decrease in the frequency factor (A in the Arrhenius equation $k = Ae^{-E/RT}$). In n-propyl halides (2 H and 1 Me) there is only one conformation for the transition state whereas for ethyl halides (3 H) there are three equivalent conformations. Thus, the frequency factor for n-propyl halides is 1/3 that for the ethyl halides, and so the rate constant (k) of the former will be 1/3 that of the latter (on the assumption that E of both reactions is the same).

In isobutyl halides the methyl groups will produce a large steric effect since at least one methyl group will be fairly close to X or Y. It has been shown experimentally that isobutyl halides are less reactive than n-propyl halides. Finally, in neopentyl halides, the presence of three methyl groups produces a very large steric effect. In the 'normal' transition state, the entering and displaced groups are collinear. This is readily possible with all the halides except possibly isobutyl halides; but it is not possible with neopentyl halides because of the presence of the three methyl groups (in the t-butyl group).

Thus in the transition state involving the neopentyl group, the Y---C---X bonds are not collinear but 'bent away' from the t-butyl group. Such a 'bent' transition state has a large compression energy and so is far more difficult to form than a 'normal' transition state. Experimental data are in agreement with these ideas, *e.g.*, Hughes *et al.* (1946) showed the following relative (S_N2) reaction rates towards the ethoxide ion at 95°C:

$$\text{Et: isoBu : Me}_3\text{CCH2 :: 1 : 0.04: } 10^{-5}$$

These very slow S_N2 reactions of neopentyl halides occur with the neopentyl group remaining intact. By changing the solvent conditions so that the mechanism becomes S_N1, the products are no longer neopentyl derivatives but rearranged products formed by a 1,2-shift.

The foregoing discussion (of both polar and steric effects) has been purely qualitative, but Ingold (1957) has considered steric effects on a quantitative basis. One reaction discussed is the following iodine exchange by the S_N2 mechanism:

$$I^- + RI \rightarrow IR + I^-$$

The iodine positions in the transition states were calculated in terms of the amount of stretching (Al) of the C–I half-bond and the deviation (Aa) of the I—C—I bond angle from 180° (*i.e.*, the amount of bending):

Table 1

R	*Δ/(Å)*	*Δa (degrees)*
Me	0.36	0.0
Et	0.37	3.8
isoPr	0.38	5.0
t-Bu	0.40	0.0
n-Pr	0.37	3.8
isoBu	0.39	5.0
neoPe	0.43	17.6

It can be seen that there is no bending in the methyl and t-butyl transition states (these are symmetrical), but all the other transition states have bent bonds and all show stretching, some more than others. The neo-compound has both the largest stretching and the largest bending, and so will be the least reactive of all, and can be expected to be much less reactive than any of the others because of the very large increase in bending. Thus, stretching and bending are both important factors in transitions states.

Ingold has also calculated steric increments of activation energies, with the energy of the methyl transition state being taken as zero. The values obtained for the reaction follow the order neoPe > isoBu > t-Bu > isoPr > n-Pr = Et > Me. For the exchange reactions Cl^- + RCI and Br^- + RBr both followed the order neoPe > t-Bu > isoBu > isoPr > n-Pr = Et > Me. The observed increments of activation energies (with MeX as zero) are all higher than the steric increments.

If this difference is attributed to the contribution by the polar effect, then it is found that the differences may be correlated with contributions of methyl groups only on the α-carbon atom; β-methyl groups do not produce any appreciable differences. Thus, the order of polar effect is t-Bu > isoPr > neoPe = isoBu = n-Pr = Et > Me. However, when Ingold calculated the frequency factors of reaction rate, he found a factor, which was neither polar nor steric, was also operating. This he called the ponderal effect because it depends on mass and is independent of bulk and distribution of any charge which the group may carry.

The addition of neutrons (which have mass but no bulk and no charge) to an alkyl group by means of isotopic substitution would produce a pure ponderal effect (cf. the replacement of H by D in the secondary isotope effect). Thus there are three factors which can affect rates of reaction: polar, steric, and ponderal.

(ii) So far, we have discussed steric effects in alkyl halides only. However, these effects may also operate in SN reactions involving other types of compounds. A very interesting example is 1-chloroapocamphane (I). Bartlett *et al.* (1938) found that this compound does not react with reagents that normally react with alkyl halides, *e.g.*, it is unaffected when refluxed with aqueous ethanolic potassium hydroxide or with ethanolic silver nitrate. As we have seen, the hydrolysis of t-butyl chloride takes place by the S_N1 mechanism. 1-Chloroapocamphane is a tertiary chloride, but since it does not ionise, the SN' mechanism is not possible.

This failure to ionise is believed to be due to the fact that the carbonium ion is flat (sp^2 hybridisation). Removal of the chloride ion from (I) would produce a positive carbon atom which cannot become planar because of the steric requirements of the bridged-ring structure. Furthermore, since the rear of the carbon atom of the C-Cl group is 'protected' by the bridge, the S_N2 mechanism is not possible. The failure to replace bromine in 1-bromotriptycene (II) is explained similarly (Bartlett *et al.*, 1939).

On the other hand, Doering *et al.* (1953) showed that (III) gave the corresponding alcohol when heated with aqueous silver nitrate at 150°C for two days, and (IV) gave the corresponding alcohol after four hours at room temperature. The reason for this behaviour (as compared with the other bridged compounds) is not certain, but it has been suggested that the extra bonds in the larger bridge in (IV) help to relieve the strain in the formation of the carbonium ion which tries to assume a planar configuration.

(iii) Steric effects also operate in the solvolysis of tertiary halides. (Solvolysis is the nucleophilic reaction in which the solvent is the nucleophilic reagent,)

$$\underset{\text{(large strain)}}{\underset{\text{tetrahedral}}{R_3C{-}X}} \xrightarrow{H_2O} \underset{\text{(small strain)}}{\underset{\text{planar; trigonal}}{R_3C^+}} + X^-$$

Brown *et al.* (1949) showed that these compounds are subject to steric acceleration. It was shown that as R increases in size, the rate of solvolysis increases. However, the larger R is, the more slowly will the carbonium ion be expected to react with the solvent molecules, and so a factor is introduced which opposes steric acceleration. Carbonium ions can undergo elimination reactions to form alkenes, and Brown *et al.* (1950) have shown that this elimination reaction increases as the R groups become larger.

The Nature of the Halogen Atom

Experimental work has shown that the nature of the halogen atom has very little effect, if any, on mechanism, but it does affect the rate of reaction. Thus it has been found that in both S_N1 and S_N2 reactions, the rate fellows the order RI > RBr > RCI. It has been suggested that a contributing factor to this order is steric strain, since the volume order of these halogen atoms is I > Br > Cl. Another contributing factor which has been suggested is that the polarisability of the C–X bond decreases in the order C–I > C–Br > C–Cl.

Experimental work has shown that many S_N2 reactions of alkyl chlorides and bromides are catalysed by the presence of iodide ions. This may be explained on the basis that the activation energies of both steps in reaction (i) are lower than the activation energy of the one-step reaction (ii):

$$\text{(i)}\ I^- + R\text{–}Cl \rightarrow I\text{–}R + Cl^- \xrightarrow{HO} ROH + I^- + Cl^-$$

$$\text{(ii)}\ HO^- + R\text{–}Cl \rightarrow HO\text{–}R + Cl^-$$

As an outcome of much experimental work, it has been found that the order of ease of displacement of groups is not fixed. In general, the stronger the displaced group is as a base, the less facile is it as a 'leaving group' in displacement reactions. However, the order of ease of displacement depends on a number of factors, *e.g.*, the nature of the alkyl group, solvent, etc. It appears that the order is generally:

$$OTs > I > Br > Cl > OH_2^+ > F > OAc > NR_3^+ > OR > NR_2$$

The Nature of the Nucleophilic Reagent

The more pronounced the nucleophilic reactivity of the reagent, the more the S_N2 mechanism will be favoured as compared with the S_N1 mechanism, since in the latter the nucleophilic reagent does not enter into the rate-determining step.

It can be anticipated that as nucleophilic reactivity decreases, the rate of an S_N2 reaction will decrease for a given series of substitutions (under similar conditions), and when the nucleophilic activity is sufficiently low, the mechanism may change from S_N2 to S_N1.

Just as the order of ease of displacement of groups depends on a number of factors, so does the order of nucleophilicity (nucleophilic reactivity). In general, the order of nucleophilicity is:

$$PhS^- > CN^- > I^- > EtO^- > OH^- > Br^- > PhO^- > Cl^- > Me_3N$$

It should be noted that nucleophilicity is the ability to form bonds to carbon atoms, whereas basicity is the affinity for protons. It should also be noted that nucleophilicity is a kinetic property, *i.e.*, is a measure of rate of reaction, whereas basicity is a thermodynamic property, *i.e.*, is a measure of the value of the equilibrium constant. Even so, it might have been expected that there would be some sort of parallelism between the two. Although this is often the case, deviations also occur, particularly with elements in the same periodic group, *e.g.*, RS^- is much more reactive as a nucleophile but is much less basic than RO^-. One possible explanation is that the former has greater polarisability than the latter. On the other hand, the degree of salvation of the two ions will be different, and this may be an important factor.

The Effect of the Solvent on Mechanisms and Reaction Rates

Experimentally, it has been found that the ionising power of a solvent depends on at least two factors, dielectric constant and salvation.

Dielectric Constant: A very rough generalisation is that ionisation of the solute increases both in amount and speed the higher the dielectric constant of the solvent.

Solvation: This factor appears to be more important than the dielectric constant. Solvation is the interaction between solvent molecules and solute molecules, and is partly accounted for by the attraction of a charge for a dipole. If the solute has polarity, then solvent molecules will be attracted to the solute molecules. The greater the polarity of the solvent, the greater the attraction and consequently the more closely the solvent molecules will be drawn to the solute molecules. Thus more electrostatic work is done and so more energy is lost by the system, which therefore becomes more stable. Hence, increasing the dielectric constant of the solvent increases the ionising potentiality of the solute molecules, and the higher the polarity of the solvent the more stable becomes the system due to increased solvation. Solvation, however, may also be partly due to certain chemical properties, *e.g.*, sulphur dioxide has an electrophilic centre (the sulphur atom carries a positive charge); hydroxylic solvents can form hydrogen bonds.

Some common solvents and their dielectric constants and dipole moments (in Debye units) are water (81.1; 2-3); formic acid (48.0; 1.5); nitrobenzene (35.7; 4.0); ethanol (25.8; 1.7); acetone (21.3; 3.0); acetic acid (7.1; 1-1.5); chloroform (4.6; 1.1); ether (4.3; 1.25); benzene (2.3; 0); carbon tetrachloride (2.2; 0).

There is also another problem that may arise. This is that although the solute molecules have ionised, the oppositely charged ions behave as a single unit, the pair being held together by electrostatic attraction. Such a complex is known as an ion-pair, and their recombination is known as internal return. It has now been shown that the majority of reactions involving carbonium-ion intermediates proceed via ion-pairs rather than dissociated ions. According to some authors there are two types of ion-pairs:

(i) **Intimate or Internal Ion-pairs:** These are enclosed in a solvent cage and the ions of the pair are not separated by solvent molecules.

(ii) **Loose or External Ion-pairs:** The ions of these pairs are separated by solvent molecules but still behave as a pair. External ion-pairs may also give rise to ion-pair return (external return), but they are more susceptible to attack by other reagents than are intimate ion-pairs.

Thus, when ionisation takes place, the following steps are possible:

$$\xrightarrow{\text{Ionisation}} \qquad\qquad \xrightarrow{\text{Dissociation}}$$

$$RX \underset{-1}{\overset{1}{\rightleftharpoons}} \underset{\substack{\text{intimate}\\\text{ion-pair}}}{R^+X^-} \underset{-2}{\overset{2}{\rightleftharpoons}} \underset{\substack{\text{external}\\\text{ion-pair}}}{R^+ \| X^-} \overset{3}{\rightleftharpoons} \underset{\substack{\text{dissociated}\\\text{ions}}}{R^+ + X^-}$$

N.B. : (i) –1 is internal return; (ii) –2 is external return; (iii) only equilibrium 3 is sensitive to a common ion effect; this is because an ion-pair behaves as a single particle, as has been shown by the effect on the depression of the freezing point (i = 1); (iv) the formation of ion-pairs is favoured in solvents of low polarity. This is due to the fact that highly polar solvents, although they assist ionisation, *i.e.*, encourage the S_N1 mechanism, also cause dissociation by solvation of the ions.

There is a great deal of evidence to support the formation of ion-pairs as intermediates, *e.g.*, the special salt effect. In the acetolysis of some alkyl tosylates, the rate of acetolysis is increased sharply on the addition of small amounts of lithium bromide or lithium perchlorate. On further addition of the salt, the rate drops to the normal essentially linear acceleration (which is caused by the normal salt effect). The explanation offered for the special salt effect is as follows.

$$\underset{(I)}{ROTs} \rightleftharpoons \underset{(II)}{R^+OTs^-} \rightleftharpoons \underset{(III)}{R^+O \| Ts^-} \rightleftharpoons \underset{(IV)}{R^+ + OTs^-}$$

$$R^+O \| Ts^- \xrightarrow{LiBr} \underset{(V)}{R^+ \| Br^-} \qquad R^+ + OTs^- \xrightarrow{LiBr} RBr$$

The solvent-separated ion-pair (III) may collapse to regenerate (II) and (I) or dissociate to give (IV). In the presence of, *e.g.*, lithium bromide, (III) but not (II) is trapped' to form (V) and so is prevented from reforming (II); (IV) reacts in the normal way to give the products. Because of the removal of (III) to form (IV) and thence the products, the return of (III) to (11) and (I) is prevented, thereby rapidly increasing the rate of formation of the product.

A point to note about intimate ion-pairs is that their geometry is very similar to that of the transition states from which they are derived. Thus, in an intimate ion-pair, the negative ion is very close to the carbonium-ion face from which it departed.

In general, S_N1 reactions will be written as direct dissociation into the two ions unless ion-pairs must be used to explain, in detail, the course of the reaction.

A number of equations have been proposed correlating rates and the nature of the solvent, but none is completely general. Hughes and Ingold (1935, 1948) proposed the following qualitative theory of solvent effects: (i) Ions and polar molecules, when dissolved in polar solvents, tend to become solvated. (ii) For a given solvent, solvation tends to increase with increasing magnitude of charge on the solute molecules or ions. (iii) For a given solute, solvation tends to increase with the increasing dipole moment of the solvent. (iv) For a given magnitude of charge, solvation decreases as the charge is spread over a larger volume. (v) The decrease in solvation due to the dispersal of charge will be less than that due to its destruction.

Since the rate-determining step in the S_N1 mechanism is ionisation, any factor assisting this ionisation will therefore facilitate S_N1 reactions. Solvents with high dipole moments are usually good ionising media and, in general, it has been found that the more polar the solvent the greater is the rate of S_N1 reactions. We have however, also to consider the problem of solvation.

$$R-X \xrightarrow{\text{slow}} \overset{\delta+}{R}---\overset{\delta-}{X} \xrightarrow{\text{fast}} R^+ + X^- \xrightarrow[\text{slow}]{OH^-} ROH$$

Increasing the polarity of the solvent will greatly increase the reaction rate, and since the transition state has a larger charge than the initial reactant molecule, the former is more solvated than the latter (rule ii). Thus the transition state is more stabilised than the reactant molecule, and solvation therefore lowers the energy of activation and so the reaction is assisted. The rates of S_N2 reactions are also affected by the polarity of the solvent.

$$HO^- \curvearrowright R-N \xrightarrow{\text{slow}} \overset{\delta-}{HO}---R---\overset{\delta-}{X} \xrightarrow{\text{fast}} HO-R+X^-$$

A solvent with high dipole moment will solvate both the reactant ion and the transition state, but more so the former than the latter, since in the latter the charge, although unchanged in magnitude (δ– = – 1/2), is more dispersed than in the former (rule iv). Thus solvation tends to stabilise the reactants more than the transition state, *i.e.*, the activation energy is increased and so the reaction is retarded.

Now let us consider the Menschutkin reaction (1890):

$$R_3\ddot{N} \curvearrowright R-N \rightarrow R_3\overset{\delta+}{N}---R---\overset{\delta-}{X} \rightarrow R_4N^+X^-$$

The charge on the transition state is greater than that on the reactant molecules; hence the former is more solvated than the latter. Thus the energy of activation is lowered and the rate of reaction thereby increased. Also, the greater the polarity of the solvent, the greater should be the solvation. The foregoing predictions have been observed experimentally.

In the following S_N2 reaction, charges decrease in the transition state,

$$HO^- \curvearrowright R\overset{+}{N}R_3 \rightarrow \overset{\delta-}{HO}---R---\overset{\delta+}{N}R_3 \rightarrow HOR + R_3N$$

and hence increasing the polarity of the solvent will retard the reaction: and retardation will be greater than that in the S_N2 hydrolysis of alkyl halides (see above; only the hydroxide ion is charged in this case).

The polarity of the solvent not only affects rates of reactions, but may also change the mechanism of a reaction, *e.g.*, Olivier (1934) showed that the alkaline hydrolysis of benzyl chloride in 50 percent aqueous acetone proceeds by both the S_N2 and S_N1 mechanisms. In water as solvent, the mechanism was changed to mainly S_N1. The dipole moment of water is greater than that of aqueous acetone, and consequently the ionisation of benzyl chloride is facilitated.

Another example we shall consider is the hydrolysis of the alkyl bromides, MeBr, EtBr, isoPrBr and t-BuBr. As we have seen, Hughes, Ingold *et al.* showed that in aqueous alkaline ethanol the mechanism changed from S_N2 for MeBr and EtBr to both S_N2 and S_N1 for isoPrBr, and to S_N1 for t-BuBr. These results were explained by the + I effects of the R groups, but it also follows that the greater the ionising power of the solvent, the less will be the + I effect of an R group necessary to change the mechanism from S_N2 to S_N1. Formic acid has been found to be an extremely powerful ionising solvent for alkyl halides, and the relative rates of hydrolysis, at 100°C, for the above series of bromides with the very weak nucleophilic reagent water, dissolved in formic acid, was found to be (Hughes *et al.*, 1937, 1940): MeBr, 1.00; EtBr, 1.71: is oPrBr, 447; t-BuBr, ca. 10^8.

This continuous increase in reaction rate shows that the mechanism is mainly S_N1 (the rate increasing with the increasing + I effect of the R group). Thus, both MeBr and EtBr are also hydrolysed by the S_N1 mechanism under these favourable conditions of high solvent-ionising power. Solvents may also affect the proportions of the products in competitive reactions, *i.e.*, the attack on the same substrate by two substituting reagents in the same solution:

$$RY \xleftarrow{Y} RX \xrightarrow{Z} RZ$$

In the S_N2 mechanism there is only one reaction step, and so the overall rate and product ratio will be determined by that stage. In the S_N1 mechanism, however, the rate is determined by the rate of ionisation of RX, and the product ratio is thus determined by the competition of the fast second steps. It therefore follows that for solvent changes, in the S_N2 mechanism the rate and product ratio will proceed in a parallel fashion, whereas in the S_N1 mechanism the rate and product ratio will be independent of each other. A simple example that illustrates this problem is the solvolysis of benzhydryl chloride (diphenylmethyl chloride). Hammett *et al.* (1937, 1938) showed that the solvolysis of benzhydryl chloride in initially neutral aqueous ethanol gave benzhydryl ethyl ether and benzhydrol. Hughes, Ingold *et al.* (1938) showed that if ethanol is first used as solvent and then water is progressively added, the overall rate increases, but there is very little increase in benzhydrol formation; the main effect is an increased rate of formation of benzhydryl ethyl ether. Hence the rate of the reaction and the ratio of the products are determined independently; this is consistent with the S_N1 mechanism but not with the S_N2.

It can be seen from this example that kinetic solvent effects may be used to differentiate between S_N2 and S_N1 mechanisms.

The above discussion of solvent effects has been purely qualitative, but it is also of interest to consider the problem in a quantitative way. This involves the use of the following thermochemical cycle.

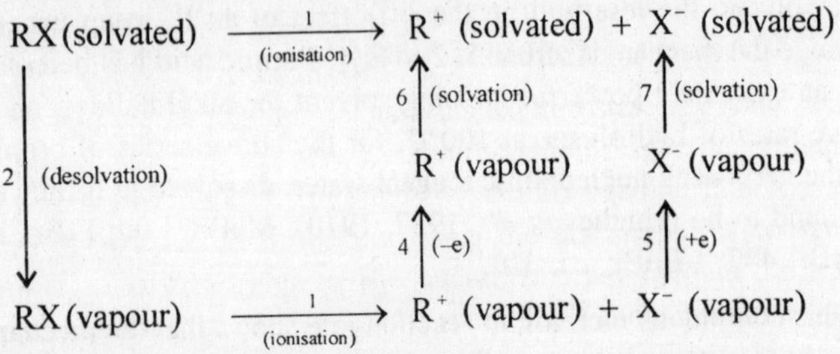

Fig. 10

Then, if the enthalpy change for each step is represented by the number (of the step) as a subscript, it follows from Hess's law that

$$\Delta H_1 = \Delta H_2 + \Delta H_3 + \Delta H_4 + \Delta H_5 + \Delta H_6 + \Delta H_7$$

(where ΔH_4 is the ionisation energy of R . and ΔH_5 is the electron affinity of X).

All of these values except ΔH_2 have been calculated from suitable experimental data, and the value of ΔH_2 has been estimated.

In the gas phase, if the reaction is homolytic, ΔH for, step 3 in the above cycle is ΔH_3. If we use methyl chloride as our example, then ΔH_3 is +80 kcal. If, however, the reaction is heterolytic,

$$\Delta H = \Delta H_3 + \Delta H_4 + \Delta H_5 = + 334.7 + 970.7 - 364.0 = 941.4 \text{ kJ}$$

It can be seen that heterolytic reaction in the gas phase is very unfavourable energetically. If the reaction is carried out in water (as solvent), then

$$\Delta H_1 = +33.7 + 334.7 + 970.7 - 364.0 - 343.1 - 288.7 = +343.1 \text{ kJ}$$

This much more favourable value is due to the large contribution of the heats of solvation (ΔH_6 and ΔH_7).

As we have already seen, in the S_N1 mechanism (with all ions and molecules solvated):

$$RX_{(aq.)} \underset{}{\overset{slow}{\rightleftharpoons}} R^+_{(aq.)} + X^-_{(aq.)} \xrightarrow[fast]{H_2O} ROH_{(aq.)} + H^+_{(aq.)}$$

the rate order is t-Bu > isoPr > Et > Me. The following table (Frazer and Singer, 1964) gives the values of AHF (heat of ionisation in kJ mol^{-1}) for various alkyl halides.

Group	*Cl*	*Br*	*I*
Me	343.1	343.1	364.0
Et	238.5	213.4	238.5
isoPr	104.6	96.23	115.1
t-Bu	71.13	62.76	75.31

Inspection of this table shows that in going in a downward direction for a given halogen atom, ΔH_1 decreases rapidly. Moreover, calculations have shown that ΔH_1 and E (activation energy) are almost identical. Hence the rate of the S_N1 reaction will increase in the downward direction, *i.e.*, t-Bu > isoPr > Et > Me.

It has been mentioned that the `rates of both S_N1 and S_N2 reactions for a given alkyl group is RI > RBr > RCI. The above table indicates the order (for S_N1) RBr > RCI > RI. This illustrates the point that even though a quantitative approach to a problem is always desirable, unless all the necessary values are known (reasonably) accurately, estimates may lead to some wrong conclusions. This is more the case with reactions in solution than with gas-phase reactions.

It has been mentioned above that a number of equations have been proposed correlating rates of solvolysis and the nature of the solvent. We shall now discuss this problem in a little more detail. One of the outstanding difficulties in this connection is that the structure of liquids is still uncertain. One quantitative correlation was proposed by Grunwald and

Winstein (1948, 1951). This is the linear free energy equation:

$$\log (k/ko) = mY$$

The standard reaction chosen for this equation is the solvolysis of t-butyl chloride, and the standard solvent is 80 per cent aqueous ethanol (80 volumes of ethanol and 20 volumes of water). This substrate was chosen for the standard reaction because it has been well established that it undergoes solvolysis by the S_N1 mechanism.

Hence, ko is the rate constant for the solvolysis of t-butyl chloride by the S_N1 mechanism in the standard solvent. k is the rate constant for the solvolysis of any particular compound in any given solvent; Y is a parameter that is a measure of the ionising power of the given solvent; m is a parameter that is a measure of the sensitivity of the rate of solvolysis of the particular compound to changes in the ionising power of the solvent. By definition, m = 1.00 for t-butyl chloride, and Y = 0.00 for the standard solvent.

Experimental results showed that the values of m were close to unity for compounds undergoing solvolysis by the SNI mechanism, but were much smaller (between 0.25 and 0.35) for compounds undergoing solvolysis by the S_N2 mechanism. This is in keeping with expectation that S_N1 reactions would be sensitive to the ionising power of the solvent, whereas S_N2 reactions would not. Thus, the determination of m could be used as a means of ascertaining whether a particular compound is undergoing solvolysis by the S_N1 or S_N2 mechanism in a given solvent. If m and Y were truly characteristic of the nature of the compound and the solvent, respectively, the plot of log k for a particular compound against the values of Y of different solvents would be a straight line with slope m.

In practice, many deviations from the equation have been observed. Winstein *et al.* (1956) modified the original equation, and other workers have also proposed various correlations. The difficulty of obtaining a good correlation between rates of solvolysis and some property of solvents is due to the fact that, apart from the problem of the structure of liquids, factors other than those considered in deriving the equations must be determined and consequently included in the equation.

4

Stereochemistry of Antimony Compounds

INTRODUCTOIN

Some optically active tervalent antimony compounds have been prepared, the phenoxastibine (I) and the stibiafluorene (II; Campbell, 1947, 1950). The asymmetry in (I) is probably due to the presence of a stable 'asymmetric' antimony atom. Campbell *et al.*, (1958) have also resolved the stibine (III).

STEREOCHEMISTRY OF SULPHUR COMPOUNDS

Various types of sulphur compounds have been obtained in optically active forms.

Sulphonium Salts

Pope and Peachey (1900) prepared carboxymethylethyl-methyl-sulphonium bromide by the reaction between ethyl methyl sulphide and bromoacetic acid. Since sulphur is tercovalent unielectrovalent in sulphonium salts (sp' hybridisation), the reaction may be formulated as

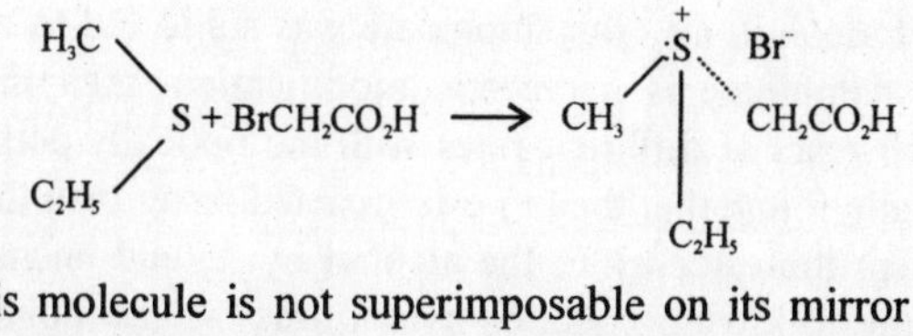

shown. This molecule is not superimposable on its mirror image, and hence can, at least theoretically, exist in two optically active forms. This bromide was treated with silver (+)-camphorsulphonate and the salt obtained was fractionally crystallised from a mixture of ethanol and ether.

Pope and Peachey found that the (+)-sulphonium camphor-sulphonate was the less soluble fraction, and had an M_D of + 68°. Since the rotation of the (+)-camphorsulphonate ion is about + 52°, this leaves + 16° as the contribution of the sulphonium ion to the total rotation. Although this does not prove conclusively that the sulphur compound is optically active, it is certainly strong evidence in its favour. Final proof was obtained by replacement of the camphorsulphonate ion by the platinichloride ion to give $[CH_3(C_2H_5)SCH_2CO_2H]_2^+ PtCl_6^=$; this compound had an $[\alpha]_D$ of +4.5° in water. In a similar way, Smiles (1900) prepared ethylmethylphenacyl-sulphonium picrate, (1), in two optically active forms, one with an $[\alpha]_D$ of +8.1° and the other –9.2°. A more recent example of an optically active sulphonium salt is one with the sulphur atom in a ring; this compound, (II), was obtained as the optically active ion with the picrate (Mann and Holliman, 1946). Optically active sulphonium salts have been prepared from optically active sulphoxides, *e.g,* (Anderson, 1971).

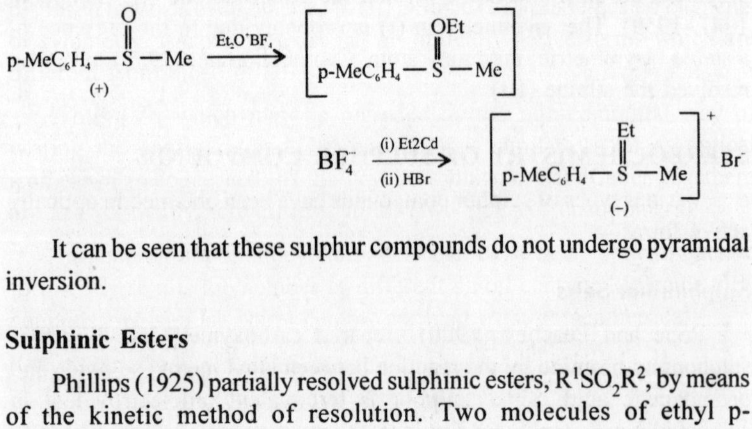

It can be seen that these sulphur compounds do not undergo pyramidal inversion.

Sulphinic Esters

Phillips (1925) partially resolved sulphinic esters, $R^1SO_2R^2$, by means of the kinetic method of resolution. Two molecules of ethyl p-toluenesulphinate were heated with one molecule of (–)-menthyl alcohol or (–)-s-octyl alcohol, *i.e.,* the sulphinate was subjected to alcoholysis. Now, if the sulphinate is a racemic modification, then the (+)- and (–)-forms will react at different rates with the optically active alcohol. Phillips actually found that the (+)-ester reacted faster than the (--)-ester. If we represent the ester by E, the alcohol by A, and unchanged ester by E_r, then the following equation symbolises the alcoholysis:

$$(+)E + (-)E + (-)A \rightarrow [(+)E(-)A] + [(-)E(-)A] + (+)E_r + (-)E_r$$

Since [(+)E(–)A] is greater than [(–)E(–)A], it therefore follows that

(+)E, is less than (–)E,; thus a partial resolution has occurred. The unchanged ester, having a lower boiling point than the new ester, distilled off first; this contained more of the (–)-form. The residual ester (the higher boiling fraction) was then heated with a large excess of ethanol; alcoholysis again occurred, this time the (–)-alcohol (menthol or octyl) being displaced to regenerate the original ethyl p-toluenesulphinate. This resulted in a fraction containing more of the (+)-form.

The optical activity of sulphinates is readily explained if their structure is (I), in which the sulphur atom is sp^3-hybridised (cf. sulphonium salts, above).

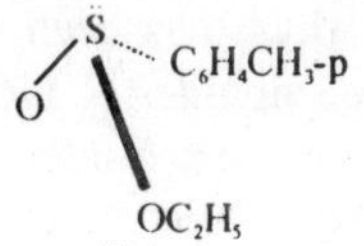

(I)

Sulphoxides

Sulphoxides of the type R^1SOR^2 have also been resolved; sulphoxides (I) and (II) were resolved by Phillips *et al.*, (1926), and Karrer *et al.*, (1951) obtained (III) in the (–)-form and the racemic modification.

Bell and Bennett (1927) investigated disulphoxides of the type

$$CH_3SOCH_2CH_2SOCH_3$$

This molecule contains two similar asymmetric sulphur atoms and so is of the type CabdCabd. Thus, it should exist in one racemic modification and one meso-form. Bell and Bennett failed to resolve this compound, but succeeded in resolving the following disulphoxide. If the former disulphoxide (the dioxide of a 1 ,4-dithian) is converted into the corresponding ring compound (*i.e.*, into a cyclic 1,4-dithian), then two geometrical isomers are possible, neither of which is resolvable; these two forms have been isolated by Bell and Bennett (1927, 1929). Shearer (1959) has examined the trans-form by X-ray analysis and showed that the ring is in the chair form with the So groups in trans and axial positions.

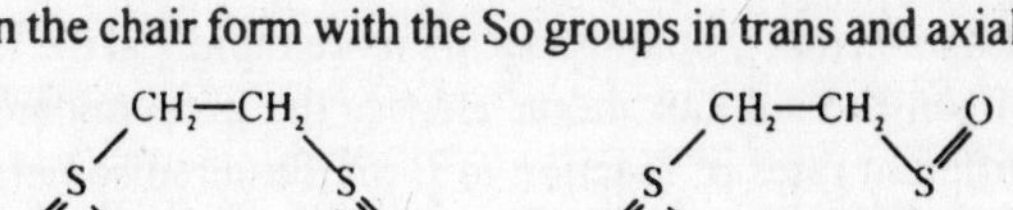

cis trans

Thianthren dioxide (IV) also exists in two geometrical isomeric forms, a, m.p. 284°C, $\mu = 1.7$ D; and β, m.p. 249°C, $\mu = 4.2$ D (Bergmann *et al.,* 1932). Hosoya *et al.,* (1957) have examined the α-form by X-ray analysis and showed it was boat-shaped (only this part of the molecule is shown in the diagrams), with the molecule folded along the S-S axis (cf. the dithian dioxides above). These authors also showed that this a-form has the anti-cis-configuration of the two S=O bonds. The -form is therefore assumed to be a trans-form.

When either of these disulphoxides is oxidised to the disulphone, both give the same compound (Hosoya, 1958). The sulphoxides described above have been obtained in optically active forms by resolution of their racemates. However, optically active sulphoxides have also been prepared by asymmetric synthesis, which has been carried out by oxidation of unsymmetrically substituted sulphides, R^1SR^2, by optically active peroxy-acids, *e.g,* α-substituted monoperoxy-glutaric acids (Balenovic *et al.,* 1960, 1961). Montanari et al. (1968) have oxidised racemic alkyl aryl sulphoxides with 0.5 molar equivalents of optically active peroxy-acids and obtained mixtures of sulphones and optically active sulphoxides. This method is based on the preferential oxidation of one enantiomer of the sulphoxide (see below). On the other hand, Savige *et al.,* (1965) have carried out an asymmetric synthesis of disulphoxide monoxides (thiosulphinates) of the type ArSOSAr by oxidation of the diaryl disulphide, with peroxycamphoric acid (Ar = Ph, p-Cl–C_6H_4, p-Me–C_6H_4, or 2-naphthyl).

Since a chiral reagent (asymmetric peroxy-acid) has been used in all of these oxidations, if we treat the two lone pairs of electrons on the sulphur as enantiotopic 'groups', then different amounts of the enantiomeric sulphoxides can be expected.

Cram *et al.,* (1963) have prepared diastereoisomeric sulphoxlaes by the oxidation of (–)-2-octyl phenyl sulphide (V) with t-butyl hydroperoxide (an achiral reagent). Using modified Newman formulae, with the sulphur in front, we can represent the reaction as shown.

Cram assumeu that oxidation occurs more rapidly at the less hindered electron pair on the sulphur atom. Hence, the predominant product is (VI). The different rates of reaction to form diastereoisomers in unequal amounts is to be expected, since the two electron pairs may be regarded as diastereotopic 'groups'. Henbest *et al.,* (1966) have shown that alkyl aryl sulphides may be stereoselectively oxidised to sulphoxides in the presence of growing aerobic cultures of Aspergillus niger (4-98 per cent optical purity).

An interesting example of an optically active sulphoxide is its preparation by transfer of a chiral centre at a carbon atom in the molecule to a sulphur atom. Mislow *et al.*, (1968) prepared (S)-α-methylallyl p-toluenesulphenate (which contains an asymmetric carbon atom) and found that it spontaneously rearranged to (S)-trans-crotyl p-tolyl sulphoxide (which contains an asymmetric sulphur atom). The rearrangement was shown to take place via a cyclic transition state. Racemisation of optically active- sulphoxides has not been studied very much, but recently, Henbest *et al.*, (1964) have shown that the sulphoxide group in certain compounds can be inverted by heating. (+)-Benzyl p-tolyl sulphoxide is racemised by heating in, *e.g*, decalin solution at 162°C. The authors also showed that cis- and trans-cyclic sulphoxides, (VIII), when heated separately in decalin at 190°C, give the same cis-trans mixture (in the ratio 1 :4). It should be noted that the t-butyl group is always equatorial and that the le,4e form (trans) is more stable than the la,4e (cis).

Mislow *et al.*, (1964) have shown that optically active sulphoxides are rapidly racemised at room temperature by solutions of hydrogen chloride in organic solvents such as benzene, dioxan, etc. The mechanism is believed to occur by nucleophilic attack of the halide ion on the protonated sulphoxide. Montanari *et al.*, (1968) have also investigated this racemisation process (and reduction of the sulphoxide to sulphide) by iodide ion in aqueous perchloric acid solution. The mechanism proposed for the racemisation is:

$$\rangle SO + H^+ \rightleftharpoons \rangle \overset{+}{S}OH \xrightarrow[\text{slow}]{I^-;\,H^+} I-\overset{+}{S}\langle \; + H_2O \overset{I^-}{\rightleftharpoons}$$

$$\rangle \overset{+}{S}-I + H_2O \longrightarrow H^+ + I^- + HO-\overset{+}{S}\langle \; \rightleftharpoons OS\langle \; + H^+$$

Also, Hammond *et al.*, (1965) have racemised sulphoxides photochemically. *e.g*, (+)-methyl p-tolyl sulphoxide, on irradiation with a mercury vapour lamp, gave the racemised product.

Johnson *et al.*, (1965) have inverted sulphoxides by chemical means The sulphoxide is treated with triethyloxonium borofluoride followed by alkali. The mechanism is uncertain, but the authors believe that inversion occurs by an S_N2 mechanism in the stage involving the hydroxide ion, *i.e.*,

$$EtO-\overset{+}{S}< + OH^- \rightarrow \overset{\delta-}{EtO}--S\langle --\overset{\delta-}{O}H \rightarrow EtO^- + >\overset{+}{S}-OH \xrightarrow{-H^+} >S=O$$

It is of interest to note, in connection with optically active sulphoxides, that Schmid and Karrer (1948) have isolated sulphoraphen from its glycoside which occurs in radish seed. These authors showed that sulphoraphen is a laevorotatory oil which owes its optical activity to the presence of a sulphoxide group.

$$CH_3SOCH{=}CHCH_2CH_2NCS$$

sulphoraphen

The absolute configurations of a number of optically active sulphoxides have now been determined by Mislow *et al.*, (1963-66). The method is based on asymmetric synthesis, the first step being the preparation of a mixture of, *e.g*, diastereoisomeric menthyl sulphinates enriched in one form (prepared by the action of (–)-menthyl alcohol on p-toluenesulphinyl chloride in the presence of pyridine at –78°C). This ester was then subjected to the Grignard reaction, and in this way the menthyloxy group is displaced by an alkyl or aryl group, the product being a mixture of enantiomers of the sulphoxide enriched in one form.

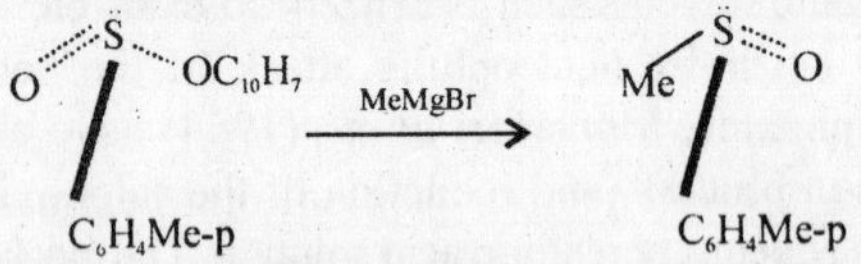

The sign of the predominant enantiomer is related to the absolute configuration of the `inducing' alcohol. Alcohols conforming to the stereoformula shown produce an excess of the (–)-(S)enantiomer. Thus terpene alcohols which have the (R)-configuration fit the stereoformula, *e.g*, (–)-menthol, (–)-borneol, etc., induce the formation of the (–)-(S)-sulphoxide. On the other hand, (+)-butan-2-ol has the (S)-configuration, and preferentially induces the formation of the (+)-(R)sulphoxide. This series of reactions therefore offers a means of configurational correlation of optically active alcohols.

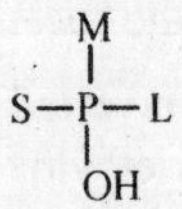

Mislow *et al.*, (1965) have also established the absolute configurations of sulphoxides by means of the ORD method.

In view of the previous discussion, it can be seen that the configuration of groups attached to sulphur in organic sulphites is also pyramidal. Pritchard *et al.*, (1968) have isolated these optically active sulphites and optically active diastereoisomers containing one asymmetric carbon atom and one asymmetric sulphur atom (cf. Cram's work, above).

One other point of interest that may be mentioned here is the stereochemistry of sulphones, $R^1SO^2R^2$. It can be seen from the formula that such compounds cannot be optically active unless the two oxygen atoms are 'different'. This has been accomplished by Stirling (1963), who synthesised (–)-benzyl p-tolyl [$^{16}O^{18}O$]sulphone ([α_D] = –0.16°). This is the first case of isotopic asymmetry for a central atom other than carbon.

Sulphilimines

Chloramine T reacts with alkyl sulphides to form sulphilimines (iminosulphuranes), *e.g,* The electronic structure of this molecule appears to be uncertain; one possibility has been given above, and in this one the sulphur atom is asymmetric (it is of the type that occurs in the sulphonium salts). An alternative electronic structure is:

In this structure, the sulphur atom can still be asymmetric. This sulphilimine has been resolved by Kenyon *et al.*, (1927). It seems likely that sulphilimines are resonance hybrids of the above two contributing structures. Lambert *et al.*, (1971) have prepared the cyclic six-membered sulphilimine, thian 1-imine, and were able to show from NMR spectroscopic studies that the parent compound is preferentially in the equatorial position, whereas its N-benzenesulphonyl and N-tosyl derivatives are preferentially in the axial position.

Sulphines

NMR spectroscopic studies have led to the conclusion that the C=S=O system in sulphines is a rigid non-linear group. Because of this, syn and anti isomers are possible.

Mangini *et al.*, (1969) have shown, from their NMR studies, that the sulphine (I)-prepared by oxidation of the thioketone with peroxy-acid-exists in two isomeric forms in about equal amounts. The authors also believe they have isolated the synp-tolyl (or the Z) form; they used column chromatography.

STEREOCHEMISTRY OF SILICON COMPOUNDS

Kipping (1907) prepared benzylethylpropylsilicyl oxide (I) and isolated one form of it. If the silicon atom has a tetrahedral configuration, this molecule is of the type CabdCabd, *i.e.*, it should exist in (+)-, (–)- and meso-forms. When, (I) was sulphonated to give (II), the latter compound was resolved. Challenger and Kipping (1910) also resolved the silane (III), and Eaborn *et al.,* (1958) have resolved the silane (IV).

More recently, Sommer *et al.,* (1964) have prepared a number of optically active silicon compounds of type (V). Many other optically active silicon compounds have been prepared by a series of chemical transformations. The relative configurations of the various silanes have been elucidated by the method of quasiracemates and by ORD measurements. The absolute configurations of 1-NpPhMeSiH and 1-NpPhMeSiF have been determined by X-ray analysis (Okaya *et al.,* 1966).

An interesting point about the chemistry of optically active silanes, $R^1R^2R^3SiZ$, is that substitution reactions (of Z) may take place with inversion or retention of configuration. Which of these occurs depends on the nature of Z and the incoming group (Y). If Z is a good leaving group and Y is more basic than Z, then reaction occurs with inversion. Hence it is possible to carry out Walden inversions with silanes, *e.g,* (1-Np = 1-naphthyl):

It is possible that reactions that occur with retention proceed via a four-centre mechanism (cf: 3 §5), *e.g,* (Attridge *et al.,* 1966):

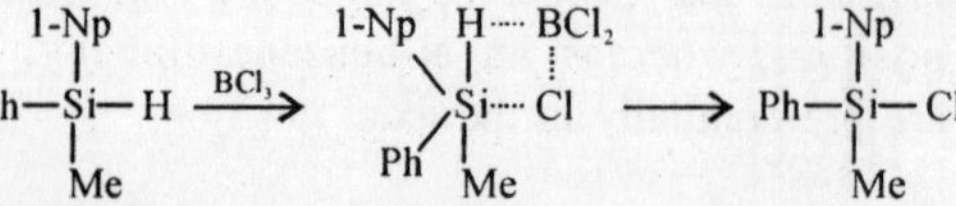

STEREOCHEMISTRY OF GERMANIUM COMPOUNDS

Schwarz and Lewinsohn (1931) obtained the (+)-form of ethylphenyl-isopropylgermanium bromide, but failed to get the (-)-form; this latter form appears to racemise in the mother liquor.

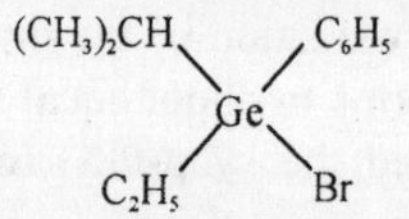

On the other hand, several workers have now synthesised a number of optically active germanium compounds by a series of chemical transformations starting from the resolved (+)- and (–)-forms of 1-NpPhMeGeH (Peddle *et al.,* 1963; Eaborn *et al.,* 1963, 1966). For Example (R = retention; I = inversion; R_3GeH = 1-NpPhMeGeH):

$$(-)\text{-}R_3GeH \xrightarrow[R]{Cl_2} (+)\text{-}R_3GeCl \xrightarrow[I]{LiAlH_4} (+)\text{-}R_3GeH$$

$$\downarrow R \;\; n\text{-BuLi}$$

$$R_3GeH \xrightarrow[R]{CO_2} (+)\text{-}R_3GeCO_2H \xrightarrow{CH_2N_2} (+)\text{-}R_3GeCO_2Me$$

Eaborn *et al.*, (1968) have also examined various reactions starting with optically active 1NpPhEtGeH, and their stereochemical assignments are based on Brewster's rules. These rules have also been applied to silicon compounds.

Stereochemistry of Selenium Compounds

Pope *et al.*, (1902) resolved carboxymethylmethylphenylselenonium bromide in the same way as the corresponding sulphonium salts; they obtained the active platinichloride (I). Mann *et al.*, (1945) also resolved selenonium salt (II). So far, attempts to resolve selenoxides have failed.

Stereochemistry of Tellurium Compounds

Lowry *et al.*, (1929) obtained the optically active forms of methylphenyl-p-tolytelluronium iodide (I), and Mann *et al.*, (1945) have resolved (II).

STEREOCHEMISTRY OF BIPHENYL COMPOUNDS

Configuration of the Biphenyl Molecule

If we assume that the benzene ring is planar, then the biphenyl molecule will consist of two planar rings; but without any further information we cannot say how these two rings are arranged spatially. Kaufler (1907) proposed the 'butterfly' formula (I) in order to account for the chemical behaviour of various biphenyl derivatives, *e.g*, Michler and Zimmermann (1881) had condensed benzidine with carbonyl chloride and obtained a product to which Kaufler assigned structure (II).

According to Kaufler, the co-axial structure (III) was impossible, since the two amino-groups are too far apart to react simultaneously with carbonyl chloride; it should be noted that this simultaneous reaction at both ends was assumed by Kaufler. Simultaneous reaction, however, is reasonable (according to Kaufler) on the folded structure (II). Now Schultz (1880) had prepared a dinitrodiphenic acid by the nitration of diphenic acid, and Schmidt *et al.*, (1903), from their work on this acid, believed it to be 6,6′-dinitrodiphenic acid (IV); these workers, it should be noted, did not synthesise the acid.

In 1921, however, Kenner *et al.*, synthesised 6,6′-dinitrodiphenic acid by means of the Ullmann reaction on the ethyl ester of 2-chloro-3-nitrobenzoic acid, and hydrolysing the product. This acid (V) (written with the two benzene rings co-axial), did not have the same melting point as Schultz's acid, and so Kenner, believing that his and Schultz's acid were both 6, 6′-dinitrodiphenic acid, suggested that the two were stereoisomers. Then Christie and Kenner (1922) showed that Kenner's acid was resolvable, and pointed out that this could be explained on the Kaufler formula (IV), since this structure has no elements of symmetry. These authors, however, also pointed out that the optical activity could also be accounted for by the co-axial structure (V), provided that the two benzene rings do not lie in one plane.

Kaufler's formula, as we have seen, was based on the assumption that the two amino-groups in benzidine react simultaneously with various reagents. Re-investigation of these reactions showed that this was not the case, *e.g,* Turner and Le Fevre (1926) found that the compound produced from benzidine and carbonyl chloride was not as originally formulated (see II or III), but had a free aminogroup, *i.e.,* the compound was

$$[NH_2C_6H,C_6H,NH]_2CO.$$

Hence Kaufler's reason for his butterfly formula is incorrect, and although it does not necessarily follow that the formula is incorrect, nevertheless Turner's work weakened Kaufler's claim. One of the strongest bits of chemical evidence for rejecting Kaufler's formula is that of Barber and Smiles (1928). These workers prepared the three dimercaptobiphenyls, (VI), (VII) and (VIII), and oxidised each one. Only one of them, the 2,2′-derivative (VI), gave the intramolecular disulphide (diphenylene disulphide, IX). On the Kaufler formula, all three dithiols would be expected to give the intramolecular disulphides, since the two thiol groups are equally distant in all three compounds.

Physico-chemical methods have also been used to determine the configuration of the biphenyl molecule, *e.g,* the crystal structure of 4,4′-biphenyl derivatives shows a centre of symmetry; this is only possible for the co-axial formula. Dipole moment measurements also confirm this configuration, *e.g,* the dipole moment of 4,4′-dichlorobiphenyl is zero; this again is only possible if the two benzene rings are co-axial.

OPTICAL ACTIVITY OF BIPHENYL COMPOUNDS

Christie and Kenner's work has been extended by other workers, who showed that compounds in which at least three of the four ortho-positions in biphenyl are occupied by certain groups could be resolved. It was then soon found that two conditions were necessary for biphenyl compounds to exhibit optical activity

(i) Neither ring must have a vertical plane of symmetry. Thus (I) is not resolvable, but (II) is.

(ii) The substituents in the ortho-positions must have a large size, *e.g,* the following compounds were resolved: 6-nitrodiphenic acid, 6,6′-dinitrodiphenic acid, 6,6′-dichlorodiphenic acid, 2,2′diamino-6,6′-dimethylbiphenyl.

The earlier work showed that three groups had to be present in the ortho-positions. This gave rise to the theory that the groups in these positions impinged on one another when free rotation was attempted, *i.e.,* the steric effect prevented free rotation. This theory of restricted rotation about the single bond joining the two benzene rings (in the co-axial formula) was suggested simultaneously in 1926 by Turner and Le Fevre, Bell and Kenyon, and Mills. Consider molecule (III) and its mirror image (IV). Provided that the groups A, B and C are large enough to 'interfere mechanically', *i.e.,* to behave as 'obstacles', then free rotation about the single bond is restricted.

Thus, the two benzene rings cannot be coplanar and consequently (IV) is not superimposable on (III), *i.e.,* (III) and (IV) are enantiomers. In molecule (III) there is no chiral centre; it is the molecule as a whole which is chiral, due to the restricted rotation.

In biphenyl the two benzene rings are co-axial, and in optically active biphenyl derivatives the rings are inclined to each other due to the steric and repulsive effects of the groups in the orthopositions. The actual angle of inclination of the two rings depends on the nature of the substituent groups, but it appears to be usually in the vicinity of 90°, *i.e.,* the rings tend to be approximately perpendicular to each other. Thus, in order to exhibit optical activity, the substituent groups in the ortho-positions must be large enough to prevent the two rings from becoming coplanar, in which case the molecule would possess a plane or a centre of symmetry, *e.g,* diphenic acid is not optically active. In configuration (V) the molecule has a plane of symmetry, and in configuration (VI) a centre of symmetry; of these two, (VI) is the more likely because of the repulsion between the two carboxyl groups.

If restricted rotation in biphenyl compounds is due entirely to the spatial effect, then theoretically we have only to calculate the size of the group in order to ascertain whether the groups will impinge and thereby give rise to optical activity. The 'size' of the group, however, must be calculated from van der Waals radii the results are in good agreement with experiment.

Later work has shown that if the substituent groups are large enough, then only two in the o- and o′-positions will produce restricted rotation, *e.g,* Lesslie and Turner (1932) resolved biphenyl-2,2'disulphonic acid (VII). In this molecule the sulphonic acid group is large enough to be impeded by the ortho-hydrogen atoms. This molecule was readily racemised on heating, but Lesslie *et al.,* (1962) prepared the enantiomers of 2,2′-di-t-butylbiphenyl from the corresponding optically active 6,6′-di-t-butylbiphenyl-3,3′-dicarboxylic acids. These were found to be highly optically stable (the t-butyl group is very large). Lesslie and Turner (1933) have also shown that the (+)-camphorsulphonate of 3′-bromobiphenyl-2-trimethylarsonium iodide (VIII) undergoes mutarotation. The trimethylarsonium group is large enough to be impeded by the ortho-hydrogen atoms (the bromine atom in the meta-position gives asymmetry to this ring). Attempts to isolate the active biphenyl compound failed because it racemised rapidly. This mutarotation indicates that the biphenyl is optically active and that the two enantiomers are readily interconvertible.

Since one phenyl group can rotate with respect to the other (about the interannular bond), the various positions would correspond to different conformations. In acyclic compounds, although there are preferred conformations the energy barriers are very low, and so the various conformations are interconvertible. In the case of the biphenyls, however, because of steric hindrance, the molecules have large energy barriers separating the two forms, and these barriers are large enough (75-105 kJ mol^{-1}) to produce separable rotational isomers. Such isomers are called atropisomers, and the two conditions for atropisomerism have been given above.

It has already been pointed out that diphenic acid is not optically active, and that its configuration is most probably (VI). Now calculation shows that the effective diameter of the carboxyl group is large enough to prevent configuration (V) from being planar, and consequently, if the two rings could be held more or less in this configuration, the molecule would not be coplanar and hence would be resolvable. Such a compound, (IX), was prepared and resolved by Adams and Kornblum (1941). The two benzene rings are not coplanar and are held fairly rigid by the large methylene bridge.

Many biphenyls have been studied from the point of view of the effect of a 2,2′-bridge on the optical activity of molecules of the type (X). When n = 1, the molecule is a disubstituted fluorene. Since this molecule is flat, it is not resolvable. When n = 2, the molecule is a disubstituted 9,10-dihydrophenanthrene. Such compounds have been resolved, *e.g,* (XVI) and (XVII) (see below). When n = 3, the molecules are resolvable and are highly optically stable. Iffland *et al.,* (1956) prepared the optically active biphenyl (XI) which has two amino-groups in the 6,6′-positions.

On the other hand, these authors have also prepared (XII) in optically active forms. Mislow (1957) has also obtained the dibenzocyclo-octadiene acids (XIII) in optically active forms; both forms were highly optically labile. Similar to (XIII) is (XIV) which has been resolved by Bell (1952). Mislow *et al.,* (1961) have also resolved the biphenyl derivative (XV), and in 1962, prepared the (–)-form of (XVI). Turner *et al.,* (1955) had prepared (XVII), and Mislow, on comparing the optical stability of (XVI) with (XVII), found that the latter was the more stable one.

In (XVI), the two methyl groups can slip past each other comparatively easily by bending out of the ring plane, but in (XVI) the second benzene ring in each naphthalene nucleus behaves as a large group and is also rigid, and consequently bending is very difficult. Mislow *et al.,* (1961) have also prepared the (–)-form of dibenzocyclononadienecarboxylic acid (XVIII); and Hall *et al.,* (1959) prepared the piperidinium salt (XIX) as the picrate and found it was optically labile.

Apart from the resolution of 2,2'-biphenyls, the stereochemistry of these compounds has been studied by u.v. and NMR spectroscopy and by X-ray analysis. In this way, a more detailed threedimensional structure of these molecules has been obtained, *e.g,* Wahl, jun. *et al.,* (1972) have carried out an X-ray diffraction study of (XX) and have shown that this molecule exists in the pseudochair form (XXI). (XX) had been prepared and resolved chromatographically (on cellulose acetate) by Lüttringhaus *et al.,* (1967).

These latter authors had, on the basis of consideration of bond angles predicted that the pseudo-tub conformation (XXII) was the more likely one. A point of interest in connection with optically active biphenyls is thatSchmidt *et al.,* (1957) have shown that 4,4′,5,5′,6,6′ hexahydroxy-diphenic acid occurs naturally in an op...cally active form.

Absolute Configurations of Biphenyls

Since biphenyls owe their asymmetry to the molecule being asymmetric, the methods used for correlating configurations of compounds containing asymmetric carbon atoms cannot be applied. However, Mislow *et al.,* (1957) used asymmetric synthesis as a means of establishing the absolute configuration of 6,6′-dinitro-2,2′-diphenic acid. Their method was chemical; assignment of absolute configuration has been obtained from a consideration of the transition states in the Meerwein-Ponndorf-Verley reduction of a dissymmetric diphenylic ketone by asymmetric alcohols of known absolute configuration. The (+)- and (–)-ketones (I) were partially reduced with (S)-(+)-methyl-t-butylmethanol in the presence of aluminium t-butoxide.

The products were unchanged ketone with the (+)-form predominating, and the (+)- and (–)-alcohols, (II), with the (–)-form predominating. Thus, as far as the unchanged ketone is concerned, this reaction is a kinetic resolution, since the ketone is now enriched in the (+)-form, and at the same time, the alcohol has become enriched in the (-)-form. Examination of models showed that, in the single conformation possible for the (S)-enantiomer (of the ketone), hydride transfer to either side of the carbonyl group is hindered by steric repulsion between the t-butyl group and a phenyl group, whereas for the (R)enantiomer, the repulsion is only between the methyl group and a phenyl group.

Thus the (+)-ketone the (+)-ketone is (S)-(+) and the (–)-alcohol is (R)-(–). Furthermore, the (S)-(+)-ketone had been prepared from (–)-6,6′-dinitro-2,2′-diphenic acid (via the dimethyl ester), and so this (–)-enantiomer is the (S)-(–)-acid. This assignment of absolute configuration has been confirmed by X-ray diffraction (Akimoto *et al.,* 1968).

The method of chemical correlation is reliable only if there is no change in configuration during the transformation or if the change in configuration occurs in a predictable manner. Thus, using the (S)-(–)-acid as absolute standard, Mislow *et al.,* carried out a number of chemical correlations, *e.g,*

Mislow *et al.,* (1958), using the (S)-(–)-acid as the absolute standard, also correlated configurations in the biphenyl series by the quasi-racemate method. In this way these authors determined the configurations of 6,6′-dichloro- and 6,6′-dimethyl-2,2′-diphenic acid. Mislow *et al.,* (1960) have also confirmed absolute configurations in the biphenyl series by the rotatory dispersion method. The authors showed that the shapes of the ORD curves depended on the configuration and conformation of the biphenyl compounds; these compounds contain inherently dissymmetric chromophores.

The specification of absolute configuration of biphenyl compounds is carried out as follows. Since biphenyls do not owe their asymmetry to the presence of asymmetric carbon atoms (which are chiral centres), the criterion now is the presence of a chiral axis (*i.e.*, an axis of asymmetry). This chiral axis may be derived from a chiral centre Z, (III), by 'extending' the point into a line AB which now passes through an elongated tetrahedron, (IV). For Z to be a chiral centre in (III), a, a′, b, and b′ must all be different, but for AB to be a chiral axis in (IV), it is sufficient that a and b be different, and a′ and b′ be different; it is not necessary that a should be different from a′, or b from b′.

To apply the sequence rule to axial chirality (or axial asymmetry), it is necessary to use an additional rule, viz. with respect to an external point on the chiral axis, groups at the near end of the axis are given precedence over groups at the far end.

If (IV) is viewed from point A, then the pair a–b, being nearer A, precedes the pair a′–b′; and in (V), pair a–b precedes pair c–d. If the order of priority is a > b and a′ > b′ (for IV), and a > b and c > d (for V), then both of these models give the final order of priority shown in (VI). If now, in accordance with the conversion rule, (VI) is viewed from the side remote from 4, then (VII) is obtained.

This, by two interchanges gives (VIIa), and since the sequence 1 → 2 → 3 is clockwise, (IV) and (V) are (R)-configurations. Had (IV) and (V) been viewed from point B on the chiral axis, then pairs a-b, a′–b′ and c–d will be 1 and 2, and the pair a–b will be 3 and 4. This gives (VIII), which in turn gives (IX), and this, by two interchanges, gives (IXa), which is still the (R)-configuration. Let us now consider some substituted biphenyls. First, the four ortho-substituent groups are inspected. If they are different, as pairs (2–6 and 2′–6′), then they are used. Thus, in (X), NO_2 = a, and CO_2H = b (priority a > h), and this moiecule is therefore the (S)-form (the last diagram has 1 → 2 → 3 anticlockwise). In (XI), since the upper ring has Cl in both ortho-positions, groups H and Me are therefore selected; NO_2 = a, CO_2H = b, Me = c, and H = d. Hence, (XI) is the (S)-form (two interchanges in the last diagram give 1 → 2 → 3 in an anticlockwise direction).

Compound (I) is the (+)-ketone described above; NO_2 = a, $-CH_2CO$ = h, and so (I) is the (S)-form.

OTHER EXAMPLES OF ATROPISOMERISM

In addition to the biphenyl compounds, there are many other examples where optical activity in the molecule is produced by restricted rotation about a single bond which may or may not be one that joins two rings. The following examples are only a few out of a very large number of compounds that have been resolved.

(i) Adams *et al.*. (1931) have resolved the following N-phenylpyrrole and N,N′-bipyrryl. Adams *et al.,* (1932) have also resolved the 3,3′-bipyridyl

(ii) 1,1′-Binaphthyl-8,8′-dicarboxylic acid has been obtained in optically active forms by Stanley (1931). This compound gives rise to asymmetric transformation; resolution with brucine gave 100 per cent of either the (+)- or (–)-compound.

Other compounds similar to the binaphthyl which have been obtained in optically active forms are 1,1′-binaphthyl-5,5′-dicarboxylic acid (I) (Bell *et al.,* 1951), the bianthryl derivatives, (II) and (III) (Bell *et al.,* 1949), and the 4,4′- and 5,5′-biquinolyls, (IV) and (V) (Crawford *et al.,* 1952).

(iii) Mills and Elliott (1928) obtained N-benzenesulphonyl-8-nitro-l-naphthylglycine (VI) in optically active forms; these were optically unstable, undergoing asymmetric transformation with brucine. Mills and Kelham (1937) also resolved N-acetyl-N-methyl p-toluidine-3-sulphonic acid (VII) with brucine, and found that it racemised slowly on standing. In both (VI) and (VII) the optical activity arises from the restricted rotation about the C–N bond (the C being the ring carbon to which the N is attached). Asymmetry arising from the same cause is also shown by 2-acetomethylamido-4′,5-dimethyldiphenylsulphone (VIII); this was partially resolved by Buchanan *et al.,* It is also interesting to note in this connection that Adams *et al.,* (1950) have isolated pairs of geometrical isomers of compounds of the types (IX) and (X); here geometrical isomerism is possible because of the restricted rotation about the C–N bonds.

(iv) Luttringhaus *et al.,* (1940, 1947) isolated two optically active forms of 4-bromogentisic acid decamethylene ether (XI). This belongs to the group known as 'ansa' compounds, and the methylene ring is perpendicular to the plane of the benzene ring; the two substituents, Br and CO_2H, prevent the rotation of the benzene nucleus inside the large ring. Cram *et al.*, (1955) have obtained paracyclophanes in optically active forms, *e.g*, (XII). In this molecule, the planes of the two benzene rings are approximately parallel (and the carboxyphenyl ring cannot rotate to give the enantiomer). When the bridges each contained four methylene groups, the compound could not be resolved (the carboxyphenyl ring can now rotate to give the enantiomers). On the other hand, Blomquist *et al.*, (1961) have resolved the simple paracyclophane (XIII).

(v) Terphenyl compounds can exhibit both geometrical and optical isomerism when suitable substituents are present to prevent free rotation about single bonds, *e.g*, Shildneck and Adams (1931) obtained (XIV) in both the cis- and trans-forms. Interference of the methyl and hydroxyl groups in the ortho-positions prevents free rotation and tends to hold the two outside rings perpendicular to the centre ring. Inspection of these formulae shows that if the centre ring does not possess a vertical plane of symmetry, then optical activity is possible. Thus, Browning and Adams (1930) prepared the dibromo cis- and trans-forms of (XV) and resolved the cis-isomer; the trans-isomer is not resolvable since it has a centre of symmetry. It can be seen from the terphenyl compounds discussed that atropisomerism does not necessarily imply enantiomerism. The different forms may be related to each other as diastereoisomers, but whether optical isomerism is also exhibited depends on the substitution pattern.

(vi) A very interesting case of restricted rotation about a single bond is afforded by the compound 10-m-aminobenzylideneanthrone (XVI). This was prepared by Ingram (1950), but he failed to resolvc it. He did show, however, that it was optically active by the mutarotation of its camphorsulphonate salt, and by the preparation of an active hydriodide. Thus the molecule is asymmetric, and this asymmetry can only be due to the restricted rotation of the phenyl group about the C-phenyl bond, the restriction being brought about by hydrogen atoms in the ortho-positions. The two hydrogen atoms labelled $\overset{\times}{H}$ overlap in space, and consequently the benzene ring cannot lie in the same plane as the 10-methyleneanthrone skeleton. Another example is the substituted cinnamic acids (XVII) (R = Cl, Me, OMe) [Adams *et al.*, 1940, 1941]. The benzene ring and the ethylenic double bond cannot become coplanar, and it was found that the order of stability to racemisation was Cl > Me > OMe.

Molecular Overcrowding

All the cases discussed so far owe their asymmetry to restricted rotation about a single bond. There is, however, another way in which steric factors may produce molecular asymmetry. It has been found that, in general, non-bonded carbon atoms cannot approach closer to each other than about 3.0Å. Thus, if the geometry of the molecule is such as to produce 'intramolecular overcrowding', the molecule becomes distorted. An example of this type is 4,5,8-trimethyl-l-phenanthrylacetic acid (I). The phenanthrene nucleus is planar and substituents lie in this plane. If, however, there are fairly large groups in positions 4 and 5, then there will not be enough room to accommodate both groups in the plane of the nucleus.

This leads to strain being produced by intramolecular overcrowding, and the strain may be relieved by the bending of the substituents out of the plane of the nucleus, or by the bending (buckling) of the aromatic rings, or by both. Thus the molecule will not be planar and consequently will be asymmetric and therefore (theoretically) resolvable. Newman *et al.*, (1940, 1947) have actually resolved it, and have also resolved (II). Bell *et al.*, (1949) resolved (III), and it was these authors who introduced the term 'intramolecular overcrowding'.

Theilacker *et al.*, (1953) resolved (IV), a heterocyclic analogue of phenanthrene. All of these compounds were found to have low optical stability, but Newman *et al.*, (1955, 1956) have prepared (V) and (VI; hexahelicene) which, so far, are the most optically stable compounds of the intramolecular overcrowding type.

It will be noticed that in (VI) the only way in which out-of-plane distortion can occur is through buckling of the molecule. The simplest molecule exhibiting overcrowding and consequent out-ofplane buckling of the molecule is 3,4-benzophenanthrene (VII); this has been shown to be nonplanar by X-ray analysis (Schmidt *et al.*, 1954). Similarly, Robertson *et al.*, (1954) have shown that (VIII) exhibits out-of-plane buckling. Another point to note in connection with out-of-plane buckling is that the buckling is distributed over all the rings in such a manner as to cause the minimum distortion in any one ring. This distortion, which enables non-bonded carbon atoms to avoid being closer together than 3.0Å (marked with dots in VII and VIII), forces some of the other carbon atoms to adopt an almost tetrahedral valency arrangement (the original hybridisation is trigonal), and this affects the physical and chemical properties of the molecule, *e.g*, Coulson *et al.*, (1955) have calculated that the deformation in (VIII) produces a loss of resonance energy of about 75 kJ mol^{-1}.

We may now summarise the problem of molecular overcrowding as follows. A molecule is said to be overcrowded if, when the standard values are assigned to the bond lengths and bond angles, at least one pair of non-bonded atoms are closer to each other than the sum of their accepted van der Waals radii. If these atoms were to remain very close to each other and the rest of the molecule remained unchanged, there would be a large steric strain.

If, however, the molecule were deformed (buckled) so that the overcrowded atoms became separated to the sum of their van der Waals radii there would now be a considerable strain energy in the molecule. Since a large amount of energy is required to stretch bonds and much less energy is required to bend them in overcrowded molecules the geometry of the molecule adjusts itself so that the energy is a minimum by mainly bending various bonds, with the bond lengths not much changed from the unstrained analogue. Thus overcrowded molecules are non-coplanar and have the form of a segment of a helix.

Mason *et al.*, (1965) have measured the circular dichroism spectra of (–)-(VIIa) and (+)-(VIIb) (these are derivatives of VII), and analysis of these spectra led the authors to conclude that (–)-(VIIa) has the M-(minus, left-handed) and (+)-(VIIb) the P-(plus, right-handed) helical configuration viewed in the direction perpendicular to the mean molecular plane. Helical molecules are those in which the arrangement of the atoms or groups is an imaginary helix. Such molecules are optically active due to the presence of helical dissymmetry or helicity. Thus, helicity is a particular type of chirality.

Just as benzene rings may suffer distortion, so can a molecule which owes its planarity to the presence of a double bond. Such an example is dianthronylidene (IX). The carbon atoms marked with dots are overcrowded (the distance between each pair is 2.9Å), and the strain is relieved by a rotation of about 40° around the olefinic double bond (Schmidt *et al.,* 1954). Even in such simple molecules as tiglic acid (X) the two methyl groups give rise to molecular overcrowding with the result that the β-methyl group appears to be displaced from the molecular plane, thereby relieving overcrowding which is also partly relieved by small distortions in bond angles. These results were obtained by Robertson *et al.,* (1959) from X-ray studies, and these authors also showed similar distortions in angelic acid (XI).

In polynuclear aromatic hydrocarbons in which the strain tends to be overcome by out-of-plane displacements of substituents and out-of-plane ring buckling, these effects cause changes in the ultraviolet spectra, but it is not yet possible to formulate any correlating rules. NMR studies by Reid (1957) have shown a shift for the hydrogen atoms in positions 4 and 5 in phenanthrene itself. A similar phenomenon has been detected by Brownstein (1958) in 2-halogenobiphenyls, and the explanation offered is that the shift is due to the steric effect between the 2-halogen and the 2′-hydrogen atom.

Although, molecular overcrowding is normally confined in the polynuclear type to systems containing three or more rings, nevertheless various substituted benzenes may also exhibit out-of-plane displacements of the substituents. Electron-diffraction studies of polyhalogenobenzenes suggest that such molecules are non-planar (Hassel *et al.,* 1947), whereas X-ray studies indicate that in the solid state such molecules are very closely or even èxactly planar (Tulinsky *et al.,* 1958; Gafner *et al.,* 1960). Ferguson *et al.,* (1959, 1961) have examined, by X-ray analysis, polysubstituted benzenes containing not more than one halogen atom, *e.g,* o-chloro- and bromobenzoic acid, and 2-chloro5-nitrobenzoic acid. In all three molecules the steric strain is relieved by small out-of-plane displacements of the exocyclic valency bonds in addition to the larger in-plane displacements of these bonds away from one another. Ferguson *et al.,* (1962) have also shown that in 2-chloro-5-nitrobenzoic acid the carboxyl group is twisted further out of the benzene plane than in o-chlorobenzoic acid.

RACEMISATION OF BIPHENYL COMPOUNDS

Since the optical activity of biphenyl compounds arises from restricted rotation, it might be expected that racemisation of these compounds would not be possible. In practice, it has been found that many optically active biphenyl compounds can be racemised under suitable conditions, *e.g*, boiling in solution. The general theory of these racemisations is that heating increases the amplitude of the vibrations of the substituent groups in the 2,2′,6,6′-positions, and also the amplitude of vibration of the two benzene rings with respect to each other, thereby permitting the substituent groups to slip by one another.

Thus the nuclei pass through a common plane and hence the probability is that the final product will contain an equimolecular amount of the (+)- and (–)-forms. Westheimer (1946-1950) has assumed, in addition to the above bond-stretchings, that the angles α, β and γ are deformed, and also the benzene rings themselves are deformed during racemisation. Westheimer obtained good agreement between the estimated and measured activation energies of racemisation of some di-orthosubstituted biphenyls. His calculations were based on known values of van der Waals radii and stretching and bending-force constants of the various bonds. In the absence of the assumption of deformations, the agreement between estimated and calculated values was very poor. At the same time, these calculations offer very strong evidence for the obstacle theory.

These ideas may be represented pictorially by the energy profile shown in Fig. 3.70. The enantiomers (A) and (A') have the same energy content, and as they approach the planar configuration (B) the energy content increases due to steric repulsion. The larger the groups a and b, the higher will be the energy barrier, *i.e.*, the slower will be the rate of racemisation.

There is, however, a further point that requires consideration. There are two possible planar configurations for the transition state, (B) and (C).

As we have seen earlier, the general principle is that crossed steric interactions between groups of different size are less than the sum of the steric interactions between groups of equal size. Hence, the transition state for racemisation is more likely to be (B) than (C).

2,2′,6,6′-Tetrasubstituted biphenyl compounds may be classified under three headings according to the nature of the substituent groups.

(i) *Non-resolvable:* These contain any of the following groups: hydrogen, methoxyl or fluorine. The volumes (effective volumes) of these groups are too small to prevent rotation about the single bond. Thus, 2,2′-difluoro-6,6′-dimethoxybiphenyl-3,3′-dicarboxylic acid (I) is non-resolvable.

(ii) *Resolvable, but Easily Raceinised:* These must contain at least two amino-groups, or two carboxyl groups, or one amino- and one carboxyl group; the remaining groups may be any of those given in (i) [but not hydrogen]. Thus, 6,6′-difluorodiphenic acid (II) is resolvable, and is readily racemised.

(iii) *Not Racemisable at All:* Biphenyl compounds which fall in this group are those which contain at least two nitro-groups; the other groups can be any of those given in (i)-but not hydrogenand (ii). Thus 2,2′-difluoro-6,6′-dinitrobiphenyl (III) is resolvable, and cannot be racemised.

The order of steric hindrance produced by various groups appears to be :

$$Br \gg Me > Cl > NO_2 > CO_2H \gg OMe > F$$

This order corresponds roughly to the order of the van der Waals radii of the groups.

In addition to the size of the groups in the ortho-positions, the nature and position of other substituent groups also play a part in the rate of racemisation, *e.g,* the rate of racemisation of (IV) is much slower than that of (V) (Adams *et al.,* 1932, 1934). Thus the nitro-group in position 3′ has a much greater stabilising influence than in position 5′. The reason for this is uncertain, but one possible explanation is as follows. In (VI), the methyl group of the methoxyl group is probably in the configuration shown. In (VII), the nitro-group in the 3′-position would tend to force the methyl group away, the resulting configuration being somewhat as shown in (VII); in this condition there would be greater interference between the methoxyl group and the two groups in the other benzene ring. This buttressing effect of groups is in the order

$$NO_2 > Br > Cl > Me$$

It is interesting to note that this order is very different from that in which these. groups produce steric hindrance in the o-position. It has also been found that groups in the 4- and/or 4′-positions affect the rate of racemisation. The reason for this is not clear. It has been proposed that the effect of a 4 - (and/or 4′-) nitro-group in retarding the racemisation of 6-nitrodiphenic acid is entirely due to a change in entropy of activation (Harris *et al.,* 1957). Adams *et al.,* (1954, 1957) have examined the rate of racemisation of (VIII). The rate is increased when R is an electron-attracting group such as NO_2 or CN, and is decreased when R is an electronreleasing group such as Me or OMe. These results were explained as follows. With, *e.g,* R = NO_2, (IX) contributes to the resonance hybrid as well as (VIII). The resonance hybrid therefore has increased C=N double bond character and consequently it is now easier for the molecule to pass through a planar transition state. With, *e.g,* R = Me, the C–N bond acquires far less double bond character than in its absence, and so it is more difficult for the molecule to pass through a planar transition state.

Adams *et al.,* (1957, 1961) also examined the optical stability of compounds of type (X); they found that the half-life was in the following order for R: Me < Et < i-Pr < t-Bu. If the effect of R were due merely to the inductive effect, then the unexpected value for t-Bu cannot be explained on this basis. The authors have proposed the following explanation. The t-Bu group, because of its large bulk, displaces the adjacent Me groups out of the plane of the benzene ring, thereby causing molecular overcrowding; this decreases the interference to rotation about the N–C (ring) bond. A molecular model of this compound showed such an interference. According to Bryan *et al.,* (1960), it is possible that steric repulsion also operates to cause considerable angle distortion.

The racemisation of 2,2′-bridged biphenyls is, in general, easy to effect, since many of them are optically labile. Furthermore, the optical stability of a bridged compound is generally considerably lower than that of the corresponding unbridged biphenyl. If, however, the bridge contains double bonds, it will have additional rigidity and so will be less optically labile, *e.g,* (XIII) is more optically stable than (X), (XI) and (XII). Racemisation of biphenyls is usually effected by heating in a suitable solvent, but Mislow *et al.,* (1963) have racemised (XIa) by irradiation (in ether solution) with ultraviolet light or by heating in the dark above 200°C.

On the other hand, the ketone (XIb) underwent simultaneous racemisatior and decarbonylation to (XIc) when irradiated under the same conditions. Mislow *et al.,* (1963) have also compared the rates of racemisation of (XII) and (XIII), and found that the deutero compound (XIII) racemised 1.13 times as fast as (XII). This is an example of the secondary isotope effect and has also been referred to as an inverse isotope effect. These results may be explained by the fact that deuterium has a smaller van der Waals radius than hydrogen, and in keeping with this explanation is that Mislow *et al.,* (1964) were unable to detect any optical activity in (XIV) (which was mixed with other deuteromethyl compounds: CD_3,CD_2H; CD_2H,CD_2H; CD_3, CDH_2).

EVIDENCE FOR THE OBSTACLE THEORY

Evidence for the obstacle theory, *i.e.,* interference of groups, amounts to proving that the two benzene rings in optically active biphenyl compounds are not coplanar. A direct chemical proof for the noncoplanar configuration was given by Meisenheimer *et al.,* (1927). The method was to unite the 'obstacle groups' in optically active biphenyl compounds, thereby forming five- or six-membered rings. Meisenheimer started with 2,2′-diamino-6,6′-dimethylbiphenyl, resolved it and then carried out the following reactions on one of the enantiomers:

In all the optically active compounds, the rings cannot be coplanar, since if they were, the molecules would possess a centre or plane of symmetry. If the dilactam, however, is not planar, then it would possess no elements of symmetry, and consequently would be optically active. If the dilactam is planar, then it has a centre of symmetry, and consequently cannot be optically active. This compound was, in fact, not optically active, and so must be planar. This planarity is readily explained in terms of resonance; all bonds have double bond character and so the molecule is planar (cf.. 2,2′-bridged biphenyls).

According to Dhar (1932), X-ray analysis studies have shown that in the solid state the biphenyl molecule is planar. On the other hand, according to Robertson (1961), who also examined crystalline biphenyl by X-ray analysis, the molecule is not strictly planar. This non-planarity has been attributed to steric repulsion between the o-hydrogen atoms. Gas phase electron-diffraction studies indicated that the two rings are inclined at about 45° to one another (Brockway *et al.,* 1944; Bastiansen, 1949). In the solid state, crystal forces presumably tend to keep the biphenyl molecule almost planar.

Ultraviolet spectra measurements of biphenyl compounds have shown that the two rings in o-substituted biphenyls are not coplanar. The ultraviolet spectrum of biphenyl, λ_{max} 248 (19 000) rim, is different from that of benzene, λ_{max} 198 (8 000) nm. This shift to the longer wavelength can be explained on the basis that biphenyl is a resonance hybrid, one of the contributing structures being the extended conjugated form (II).

Thus the interannular bond will have some partial double bond character and consequently the molecule will tend to be planar. When o-positions are occupied, the coplanarity is prevented and consequently the spectrum will be different, the maximum absorption shifting to shorter wavelengths. Thus, 2-methylbiphenyl has λ_{max} 236 (10 000) nm and 2,2′-

dimethylbiphenyl Amax 224 (~700) nm. Also, Pickett *et al.*, (1936) have shown that the ultraviolet absorption spectra of bimesityl (III),λ_{max} 267 (545) nm, and mesitylene (IV), λ_{max} 266 (260) nm, were almost identical but different from the spectrum of biphenyl. In (III) there is steric inhibition of resonance, and so coplanarity is prevented. Thus E for (III) is approximately double that of (IV), but 'Lax remains essentially unchanged.

In 2,2′-bridged biphenyls in which the bridge is saturated it is possible to calculate the angle of twist between the two benzene rings from normal bond angles and bond lengths. These values depend on the type of bridge, *i.e.*, the number of atoms and their nature (C, S, O. etc.). The ultraviolet spectra of these compounds have confirmed that the two benzene rings are not coplanar (when the bridge is of the appropriate type).

Other evidence for the obstacle theory is the behaviour of Chichibabin's hydrocarbon. Wheland *et al.*, (1952) examined this compound by means of the ESR method and found that it existed in the form of the free diradical to the extent of 4-5 per cent. When, however, there were four chlorine atoms in the four o-positions, the amount of free diradical was very much greater. In the latter, the coplanarity of the two rings is now prevented. Westheimer's calculations of the activation energies of racemisation of some biphenyl compounds are also evidence for the obstacle theory.

STEREOCHEMISTRY OF THE ALLENES

Allenes are compounds which have the general structure (I).

ahC=C=Cde   abC=C=Cah

(I)   (II)

Examination of the space formula of compounds of this type shows that the molecule and its mirror image are not superimposable. The σ–π way of writing (I). The two end carbon atoms are in a state of trigonal hybridisation, and the centre carbon atom is in the digonal state. Thus, the centre carbon atom forms two π-bonds which are perpendicular to each other; π_x-bond is perpendicular to the plane of the paper, and the π_y-bond is in the plane of the paper. In the trigonal state, the π-bond is perpendicular to the plane containing the three σ-bonds consequently the groups a and b lie in the plane of the paper, and the groups d and e in the plane perpendicular to the plane of the paper. This molecule does not possess a plane or centre of symmetry: this is also true for molecule (II). Thus (I) and (II) will be resolvable.

The resolvability of allenes was predicted by van't Hoff in 1875, but experimental verification was not obtained until 1935, when Mills and Maitland carried out a catalytic asymmetric dehydration on 1,3-di-l-naphthyl-l,3-diphenylprop-2-enol (III), to give the dinaphthyl-diphenylallene (IV). When the dehydration was carried out with an optically inactive dehydrating catalyst, *e.g,* p-toluenesulphonic acid, the racemic modification of the allene derivative was obtained.

When, however, the alcohol (III) was boiled with 1 per cent benzene solution of (+)-camphorsulphonic acid, a dextrorotatory allene was obtained. Similarly, (–)-camphorsulphonic acid gave a laevorotatory allene. Another asymmetric synthesis of an allene is that of 1,3-diphenylallene. Jacobs *et al.,* (1957) rearranged 1,3-diphenylpropyne by adsorption on alumina impregnated with brucine or quinine; the former gave the (-)-allene, and the latter the (+)-allene (cf. 2 §10vi):

$$PhC \equiv CCH_2Ph \xrightarrow[\text{alkaloid}]{Al_2O_3} PhCH{=}C{=}CHPh$$

The first successful resolution of an allene derivative was carried out by Kohler *et al.,* also in 1935. Lapworth and Wechsler (1910) prepared 3-1-naphthyl-l,3-diphenylallene-l-carboxylic acid (V) but failed

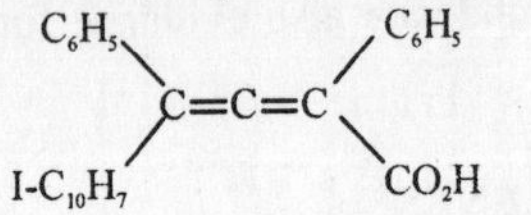

(V)

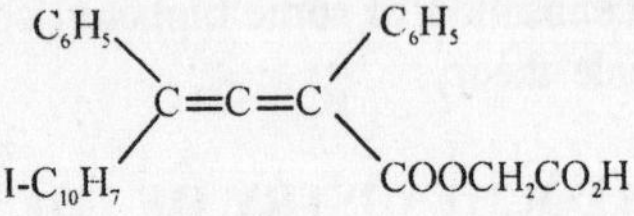

(VI)

to resolve it; they were unable to crystallise the salts with active bases. Kohler converted this acid into the glycollic acid ester (VI) and was then able to resolve (VI) by means of brucine. Wotiz *et al.,* (1951) have also resolved the simpler allenic acid (VII) by means of strychnine.

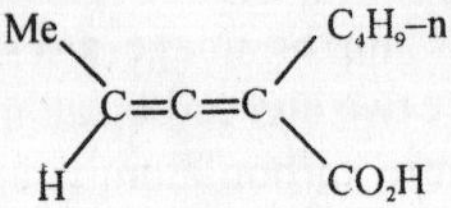

(VII)

Landor *et al.,* (1959) have prepared an optically active allene by a method which correlates it stereochemically with a tetrahedrally asymmetric alcohol. An optically active acetylenic alcohol, on treatment with thionyl chloride, gave an optically active allene; the mechanism is possibly S_Ni'.

Landor *et al.*, (1962) have also deduced the absolute configuration of the (4)-chloride by first determining the absolute configuration of the (+)-alcohol; the (R)-(–)-alcohol gave the (S)-(–)allene.

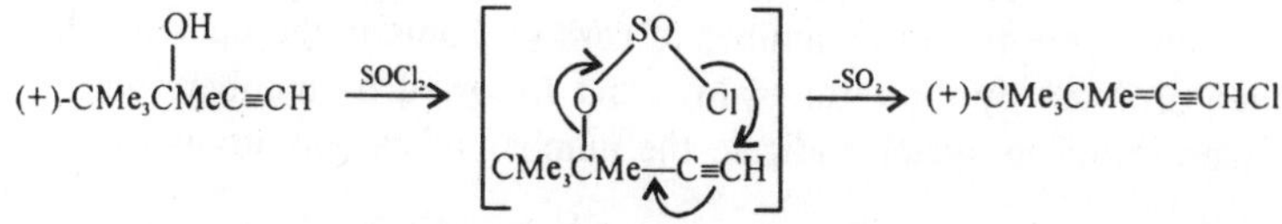

It has been previously pointed out that if the number of double bonds in the cumulene is odd the molecule exhibits geometrical isomerism, but if even, then it exhibits optical isomerism. The allenes discussed above contain two double bonds, but more recently, Nakagawa *et al.*, (1961) have prepared the following cumulene with four double bonds, 1,5-di-t-butyl-l,5-dip-chlorophenylpentatetraene, in optically active forms.

Although allenes were not successfully resolved until 1935, compounds with a similar configuration were resolved as early as 1909. In this year, Pope *et al.*, resolved 4-methylcyclohexylidene-1-acetic acid (VIII); in this compound one of the double bonds of allene has been replaced by a six-membered ring, and the general shape of the allene molecule is retained. It is interesting to note, in connection with allenes, that the antibiotic mycomycin and other natural poly-acetylenes have been shown to contain the allene grouping. Mycomycin is optically active, and owes its optical activity to the presence of this grouping. Celmer and Solomons (1953) have shown that the structure of mycomycin is:

$$CH\equiv CC\equiv CCH=C=CHCH=CHCH=CHCH_2CO_2H$$

Specification of Absolute Configuration of Allenes

This is carried out in a similar manner to that used for biphenyls. For allenes, the chiral axis passes through the double bonds.

The absolute configurations of allenes have been determined by conversion of an optically active molecule of known absolute configuration into an allene (*e.g.* Landor, above), or by converting a chiral allene into a molecule of known absolute configuration by stereochemically unambiguous reactions. The exception to the above procedures is the work of Mason, who determined the absolute configuration of 1,3-diphenylallene by means of electronic absorption and circular dichroism spectra. Lowe (1965) has now developed a method for predicting the sign of rotation of allenes of known absolute configuration.

STEREOCHEMISTRY OF THE SPIRANS

If both double bonds in allene are replaced by ring systems, the resulting molecules are spirans. One method of naming spirans obtains the root name from the number of carbon atoms in the nucleus; this is then prefixed by the term 'spiro', and followed by numbers placed in square brackets which indicate the number of carbon atoms joined to

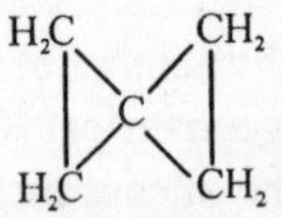

(I)

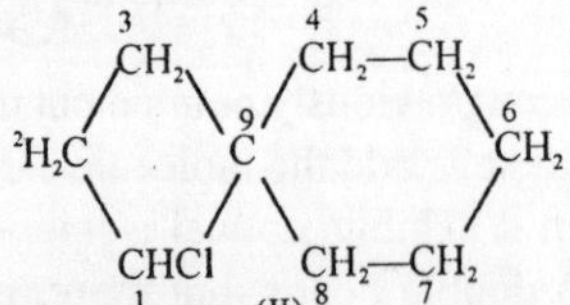

(II)

the 'junction' carbon atom. The positions of substituents are indicated by numbers, the numbering beginning with the smaller ring and ending on the junction carbon atom; *e.g,* (I) is spiro-[2,2]-pentane, (II) is 1-chlorospiro-[5,3]-nonane.

Examination of these formulae shows that the two rings are perpendicular to each other, and hence suitable substitution will produce molecules with no elements of symmetry, thereby giving rise to optically active forms, *e.g,* Mills and Nodder (1920, 1921) resolved the dilactone of benzophenone 2, 2′, 4, 4′-tetracarboxylic acid (III). In this molecule the two shaded portions are perpendicular to each other, and consequently there are no elements of symmetry. When this compound is treated with sodium hydroxide, the lactone rings are opened to form (IV) and the optical rotation disappears. Boescken *et al.,* (1928) condensed penta-erythritol with pyruvic acid and obtained the spirocompound (V), which they resolved. Some other spiro-compounds that have been resolved are the spiro-heptane (VI) (Backer *et al.,* 1928, 1929), the spiro-hydantoin (VII) (Pope and Whitworth, 1931), and the spiroheptane (VIII) (Jansen and Pope, 1932).

$$2\ CH_3COCO_2H + C(CH_2OH)_4 \rightarrow$$

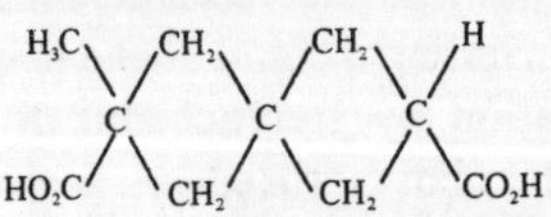

(VI)

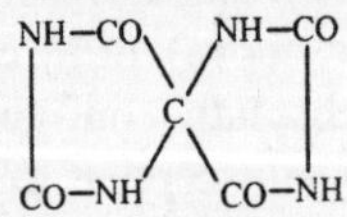

(VII)

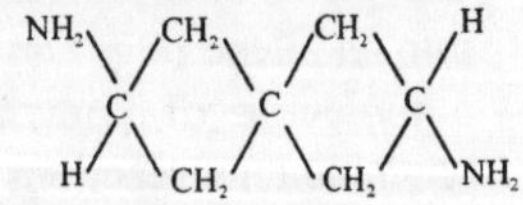

(VIII)

In all the cases so far discussed, the optical activity of the spiran is due to the asymmetry of the molecule as a whole; there is only one pair of enantiomers. If a spiro-compound also contains asymmetric carbon atoms, then the number of optically active forms is increased (above two), the actual number depending on the compound in question, *e.g,* Sutter and Wijkman (1935) prepared the spiro-compound (IX), which contains two similar asymmetric carbon atoms (o). If we imagine the left-hand ring of (IX) to be horizontal, then the righthand ring will be vertical; and if we represent them by bold horizontal and vertical lines, respectively, then there are three different geometrical isomers possible, (X), (XI) and (XII) (this can be readily demonstrated by means

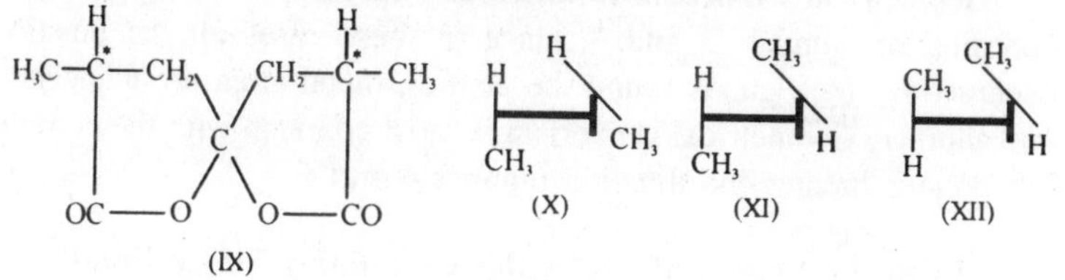

Fig. 1

of models). Each of these geometrical isomers has no elements of symmetry, and so each can exist as a pair of enantiomers. Three racemic modifications were actually isolated by Sutter and Wijkman, but were not resolved.

Cram *et al.,* (1954) have also prepared the following three spiro-[4,4]-nonanediols (as racemates): Various spiro-compounds have been prepared in which the spiro-atom is nitrogen, phosphorus or arsenic. A spiran compound, acorone, has now been found in nature. The method of specifying the absolute configuration of spirans is similar to that for allenes *e.g,*

STEREO-ISOMERISM

Stereo isomerism is exhibited by those compounds which have the same position of atoms or groups but these atoms or groups have different arrangement round the central atom.

The compounds showing stereo isomerism are called stereo isomers. Stereo isomerism may be of two types viz. geometrical (or cis-*trans*) *isomerism and optical* (or d–l or *mirror-image*) isomerism.

GEOMETRICAL (or cis-trans) ISOMERISM

The complex compounds which have the same ligands in the coordination sphere but the relative position of the ligands round the central metal atom is different are called *geometrical isomers* and the phenomenon is called *geometrical isomerism.*

In a given complex compound the two ligands may occupy positions either adjacent to each other or opposite to each other. The complex compound having two ligands occupying the adjacent positions to each other is called *cis-isomer* while that in which the two ligands occupy opposite positions is called trans-isomer. Thus geometrical isomerism is also called *cis-trans isomerism.*

Geometrical isomerism is not found in complex compound with coordination number 2 and 3, since in these cases all the positions occupied by the ligands round the central, metal atom are adjacent to one another. Geometrical isomerism is most common with the complex compounds having coordination number 4 and 6.

Table 1 : Shapes of Molecules Containing Single Bonds

Number of electrons in valency shell	*Number of bonding pairs*	*Number of lone-pairs*	*Hybrid orbitals used*	*Shape of molecule*	*Examples*
2	2	0	sp	Linear	$HgCl_2$
3	3	0	sp^2	Triangular plane	BCl_3
4	4	0	sp^3	Tetrahedron	CH_4
	3	1	sp^3	Trigonal pyramid	NH_3
	2	2	sp^3	V-shape	H_2O
5	5	0	sp^3d	Trigonal bipyramid	PCl_5
6	6	0	sp^3d^2	Octahedron	SF_6

NITROGEN COMPOUNDS

According to the electronic theory of valency, nitrogen can be tercovalent or quadricovalent unielectrovalent; in both of these states nitrogen, as the 'central' atom, can exhibit optical activity.

Quaternary Ammonium Salts

Originally, the valency of nitrogen in quaternary ammonium salts was believed to be quinquevalent; later, however, it was shown that one valency was different from the other four. Thus, using the formula, $[Nabcd]^+X^-$, for quaternary ammonium salts, and assuming that the charge on the nitrogen atom has no effect on the configuration of the cation, the cation may be considered as a five-point system similar to that of carbon in compounds of the type Cabde. This similarity is based on the assumption that the four valencies in the ammonium ion are equivalent, and this assumption is well substantiated experimentally and also theoretically.

Hence there are three possible configurations for the cation [Nabcd]+, (I), (II) and (III). If the cation is planar (I), then it would not be resolvable; it would be resolvable, however, if the configuration is pyramidal (II) or tetrahedral (III). Le Bel (1891) claimed to have partially resolved isobutylethylmethylpropylammonium chloride, (IV), by means of Penicillium glaucum but later work apparently showed this was wrong. The first definite resolution of a quaternary ammonium salt was that of Pope and Peachey (1899), who resolved allylbenzylmethylphenylammonium iodide, (V) by means of (+)-bromocamphorsulphonic acid. This was the first case of optical activity due to a 'central' atom other than carbon. This resolution was then followed by the work of Jones (1905), who resolved benzylethylmethylphenylammonium iodide.

$$\left[CH_3CH_2CH_2-\overset{\displaystyle CH_3}{\underset{\displaystyle C_2H_5}{N}}-CH_2CH(CH_3)_2\right]^+ Cl^- \qquad \left[CH_2=CHCH_2-\overset{\displaystyle CH_3}{\underset{\displaystyle C_2H_5}{N}}-CH_2C_6H_5\right]^+ I^-$$

(IV) (V)

Thus, the ammonium ion cannot be planar, but must be either pyramidal or tetrahedral. Bischoff (1890) had proposed a pyramidal structure, and this configuration was supported by Jones (1905) and Jones and Dunlop (1912). On the other hand, Werner (1911) had suggested the tetrahedral configuration, and this was supported by Neagi (1919) and Mills and Warren (1925). It was, however, Mills and Warren who gave the most conclusive evidence that the configuration is tetrahedral. Their evidence is based on the following argument.

Compounds of the type abC=C=Cab are resolvable since carbon is 'tetrahedral' and if nitrogen is also 'tetrahedral', then the compound abC=N=Cub should be resolvable, but will not be resolvable if the nitrogen is pyramidal. Mills and Warren prepared 4-carbethoxy-4'-phenylbispiperidinium-1,1'-spiran bromide, and resolved it. If the configuration of this molecule is (VI), *i.e.*, a spiran, then it possesses no elements of symmetry, and hence will be resolvable; if the configuration is (VII) (*i.e.*, pyramidal), then it will possess a vertical plane of symmetry, and hence will be optically inactive. Since the compound was resolved, the configuration must be tetrahedral, *i.e.*, (VI). This tetrahedral configuration has been confirmed by physicochemical studies. Later,

Hanby and Rydon (1945) have shown that the diquaternary salts of dimethylpiperazine exhibit geometrical isomerism, and this is readily explained on the tetrahedral configuration of the four nitrogen valencies.

Further support for the tetrahedral configuration comes from the X-ray analysis of crystalline tetramethylammonium chloride; each nitrogen atom has a tetrahedral arrangement of four methyl groups around it. Thus, quadrivalent nitrogen ($2p^3$) ($3s$) is tetrahedrally hybridised (sp^3).

It has already been mentioned that McCasland and Proskow (1956) prepared a spironitrogen compound which contained no plane or centre of symmetry, but was nevertheless optically inactive because it contained an alternating axis of symmetry. We shall now examine this compound (VIII; Y^- is the p-toluenesulphonate ion) in more detail. This molecule can exist in four diastereoisomeric forms, three active and one meso. All four have been prepared, and are depicted as shown in (IX), (X), (XI) and (XII).

The co-axis of each spiran is assumed to be perpendicular to the plane of the paper, and the intersecting lines represent the two rings. The short appendages show whether the two substituents (methyl) are cis or trans. The ring nearer the observer's eye is indicated by the heavy line and a uniform orientation has been adopted: the front ring is always vertical, and the back horizontal ring with at least one substituent directed upwards and the cis ring placed at the back in the case of the cis/traps ring combination. Racemisation of optically active quaternary ammonium salts is far more readily effected than that of carbon compounds containing a chiral centre, *i.e.,* compounds of the type Cabde. The mechanism of the racemi-sation of the ammonium salts is believed to take place by dissociation into the amine, which then rapidly racemises.

$$\text{Nabcd}\}^{+}X^{-} \rightleftharpoons \text{Nabc} + \text{dX}$$

Recombination of the racemised amine with dX results in the racemisation of the quaternary compound. This is in keeping with the fact that quaternary ammonium sulphates and nitrates are difficult to racemise. These anions are poor nucleophiles and so formation of dX ($X = \bar{O}SO_2H, \bar{O}NO_2$) is very slow.

Tertiary Amine Oxides

In tertiary amine oxides, abcNO, the nitrogen atom is joined to four different groups, and on the basis that the configuration is tetrahedral, such compounds should be resolvable. In 1908, Meisenheimer resolved ethylmethylphenylamine oxide (I) and this was then followed by the resolution of other amine oxides, *e.g,* ethylmethyl-l-naphthylamine oxide (II) and kairoline oxide (III). Bennett and Glynn (1950) have obtained two geometrical isomers of 1,4-diphenylpiperazine dioxide; this is readily explained on the tetrahedral configuration of nitrogen.

Amines

If the tertiary amine molecule, Nabc, is planar, it will be superimposable on its mirror image, and therefore cannot be optically active. All attempts to obtain tertiary amines in optically active forms have failed up to the present time, *e.g,* Kipping and Salway (1904) treated secondary amines, R^1NHR^2, with (±)-benzylmethylacetyl chloride; if the three valencies of the nitrogen atom are not planar, then the base will be a racemic modification, and on reaction with the acid chloride, the following four substituted amides should be formed: B_+A_+, B_-A_-, B_+A_-, B_-A_+, *i.e.*, a mixture of two pairs of enantiomers.

Experiments carried out with, *e.g,* methylaniline and benzylaniline gave homogeneous products. Meisenheimer *et al.*, (1924) attempted to resolve N-phenyl-N p-tolylanthranilic acid (I) and also failed. In view of these failures, it would thus appear that the tertiary amine molecule is planar. Physico-chemical methods, *e.g,* dipole moment measurements, infrared absorption spectra studies, etc., have, however, shown conclusively that the configuration of ammonia and of tertiary amines is tetrahedral. Thus ammonia has been shown to have a dipole moment of 1.5 D; had the molecule been planar, the dipole moment would have been zero. Furthermore, the nitrogen valency angles in, *e.g,* trimethylamine have been found to be 108°, thus again showing that the amine molecule is not planar.

Why, then, cannot tertiary amines be resolved? Is it a question of experimental technique, or is there something inherent in the tertiary amine molecule that makes it impossible to be resolved? Meisenheimer (1924) explained the failure to resolve as follows. In the tertiary amine molecule, the nitrogen atom oscillates rapidly at right angles above and below the plane containing the groups, (II) and (III) are the two extreme forms, and they are mirror images and not superimposable ((IV) is (III) 'turned over', and it can be seen that (IV) is the mirror image of (II)). Thus this oscillation brings about very rapid optical inversion.

This oscillation theory is supported by evidence obtained from the absorption spectrum of ammonia (Barker, 1929; Badger, 1930), and the frequency of the oscillation (and therefore the inversion) has been calculated to be 2.3×10^{10} per second (Cleeton *et al.,* 1934). This inversion of amines is best represented as an 'umbrella' switch of bonds, *i.e.,* the bond lengths remain unaltered and only the nitrogen valency angles change. This interpretation is more in keeping with the facts, *e.g,* as the groups a, b and c increase in weight, the frequency of the inversion of the molecule decreases. Theoretical calculations have shown that an optically active compound will not racemise spontaneously provided that the energy of activation for the change of one enantiomer into the other is greater than 50-63 kJ mol^{-1}. The two forms, (II) and (III), have been shown to be separated by an energy barrier of about 25 kJ mol^{-1}, and consequently the two forms are readily interconvertible.

In view of what has been said above, it would appear that tertiary amines of the type Nabc could never be resolved. Now, Kincaid and Henriques (1940), on the basis of calculations of the energy of activation required for the inversion of the amine molecule, arrived at the conclusion that tertiary' amines are incapable of resolution because of the ease of racemisation, but if the nitrogen atom formed a part of a ring system, then the compound would be sufficiently optically stable to be isolated.

This prediction was confirmed by Prelog and Wieland (1944), who resolved Troger's base (V) by chromatographic adsorption on D-lactose. In this compound, the nitrogen is tervalent, but the frequency of oscillation has been brought to zero by having the three valencies of nitrogen as part of the ring system (see also below). As a result of their CD studies, Mason *et al.,* (1967) believe that Troger's base has predominantly the folded structure (V) and that this is the (+)-isomer and has the (IR, 3R) configuration. Roberts *et al.,* (1958) have examined N-substituted aziridines (ethyleneimines) by NMR spectroscopy. Their results support the 'umbrella' switch of bonds, and these authors believe that optical resolution of this type of compounds may be possible below –50°C. N-Ethylaziridine (VI) at room temperature showed two chemical shifts for the protons, one being due to hydrogens cis and the other to hydrogens trans to the ethyl group.

At 110°C only one broad band was observed. At this temperature, the rate of inversion is fast enough for the protons to now exhibit a 'single environment'.

The rate of inversion in N-substituted aziridines is affected both by the nature of the N-substituent and by the presence of substituents attached to the ring-carbon atoms. Thus, *e.g,* in N-t-butylaziridine, the inversion rate is accelerated, whereas in N-chloro-2-methylaziridine the inversion rate is sufficiently slow for the two invertomers to be separated (Brois, 1967, 1968).

From the foregoing discussion it can be seen that the rate of inversion of nitrogen in aziridines can be slowed down considerably. If it could be slowed down completely, then it might be possible to obtain a chiral compound whose chirality is due to the tervalent nitrogen atom (cf. Troger's base). Montanari *et al.,* (1968) have prepared such a compound, 2-methyl-3,3-diphenyloxaziridine (VII) by an asymmetric synthesis, *e.g,* N-diphenylmethylenemethylamine (VIII) has been oxidised with (IS)-(+)-peroxycamphoric acid to give (–)-(VII).

(VIII) (–)-(VII)

In general, the inversion rate in solution of amines of the type R_3N is too fast to be measured by NMR spectroscopy. Saunders *et al.,* (1963), however, examined the inversion rate of N,N-dibenzylmethylamine in aqueous hydrochloric acid, basing their method on the assumption that a protonated tertiary amine cannot undergo inversion. In acid solution, protonation and deprotonation are extremely fast and consequently the averaged rate of inversion of amine molecule is reduced sufficiently for the inversion rate to be measured.

A point worth recalling here is that resolution has been carried out with substituted amines of the type $ArNR^1R^2$, where Ar is an aromatic nucleus containing at least one ortho-substituent and R^1 and R^2 are different groups (see (VI)-(VIII)). The optical activity of these compounds is not due to the asymmetry of the nitrogen atom; it is due to the asymmetry of the molecule as a whole, arising from restricted rotation and about the N-C (aryl) bond. Although the nitrogen atom is in a state of oscillation, there is always restricted rotation; the relative positions of Ar, R^1 and R^2 remain unchanged throughout these oscillations.

Tertiary amines are an example of a group of compounds in which the central atom is bonded to three groups in a pyramidal geometry and possesses one lone pair of electrons, *i.e.,* tertiary amines are of the type :MY_3. Such compounds, in which the central atom belongs to Groups IV to VI of the Periodic Table, may undergo spontaneous inversion of configuration-pyramidal (atomic) inversion -the process involving a transition state in which the bonds (from the central atom) are sp^2 hybridised and—the lone pair has pure p-character. Pyramidal inversion has been observed for nitrogen, phosphorus, and arsenic. In analogous sulphur compounds, however, because of the configurational stability of the sulphur atom in these compounds, optical isomers have been isolated. Even so, pyramidal inversion has been observed in certain cases, *e.g,* $Me_3\ddot{S}^+$.

Carbanions, *i.e.,* species of the type : CY_3^-, are also an example of pyramidal inversion.

Oximes

In 1883, Goldschmidt found that benzil dioxime,

$$C_6H_5C(=NOH)C(=NOH)C_6H_5$$

could be converted into an isomeric form by boiling it in ethanolic solution; and then, in 1889, Meyer *et al.,* isolated a third isomer of this compound. Beckmann, also in 1889, found that benzaldoxime existed in two isomeric forms, and from that time many aromatic oximes were shown to exist in two isomeric forms. The existence of isomerism in aromatic oximes was first explained by structural isomerism, two of the following four structures corresponding to the two isomers (where R is an alkyl or an aryl group). Hantzsh and Werner (1890), however, suggested that the isomerism of the oximes was geometrical and not structural.

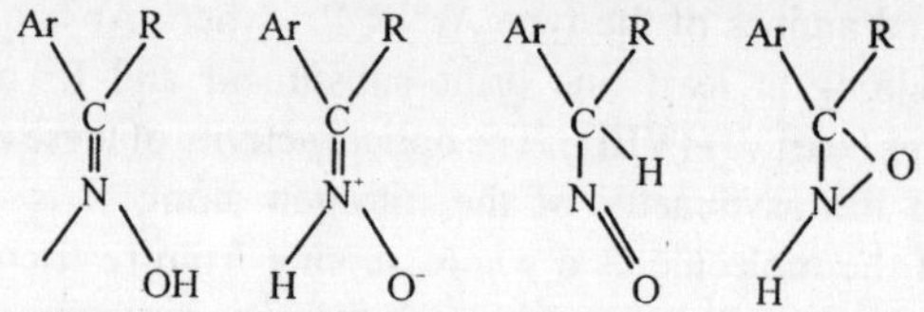

oxime (I) nitrone (II) (III) (IV)

According to these authors, nitrogen is tervalent (in oximes), and is situated at one corner of a tetrahedron with its three valencies directed towards the other three corners; consequently the three valencies are not coplanar (see also below). These authors also assumed that there is no free rotation about the C=N double bond, and therefore proposed configurations (V) and (VI) for the two isomers:

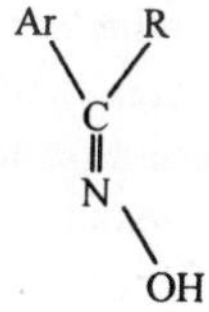

(V)

(VI)

Many facts are in favour of geometrical isomerism, *e.g,*

(i) If Ar = R, then isomerism disappears.

(ii) (III) and (IV) would be optically active; this is not found to be so in practice.

(iii) Absorption spectra measurements show that the two isomers have identical structures.

As pointed out above, Hantzsch and Werner chose structure (I) as the formula for the oximes, but examination of (II) shows that this would also satisfy the requirements for geometrical isomerism; structure (I) was chosen because oximes were known to contain the group > C=NOH. Later work, however, has shown that the problem is not so simple as this; methylation of an oxime (with methyl sulphate) usually produces a mixture of two compounds, one of which is the o-methyl ether, (VII), and the other the N-methyl ether, (VIII). These two are readily distinguished by the fact that on heating with hydriodic acid, (VII) gives methyl iodide, whereas (VIII) gives methylamine. Thus, Semper and

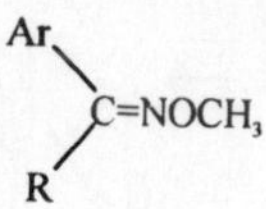

(VII)

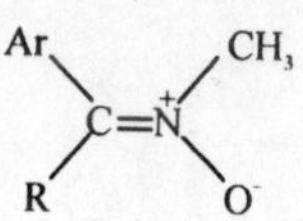

(VIII)

Lichtenstadt (1918) obtained four methyl derivatives of phenyl p-tolyl ketoxime, (IX)-(XII). On treatment with concentrated hydriodic acid, two of these compounds gave methyl iodide, and therefore correspond to the 0-methyl derivatives, (IX) and (X); the other two compounds gave

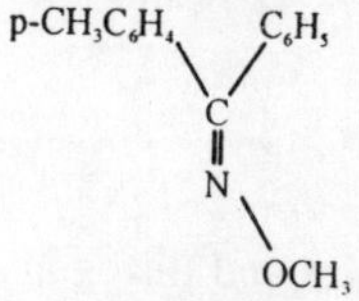

(IX)

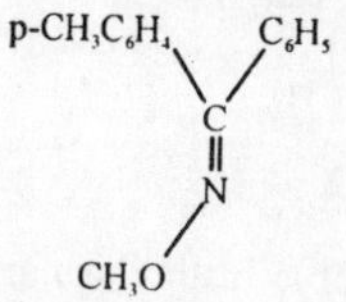

(X)

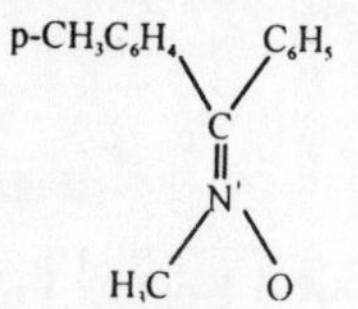

(XI)

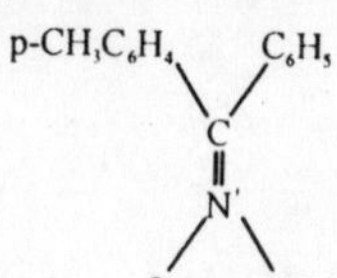

(XII)

methylamine, and therefore correspond to the N-methyl derivatives, (XI) and (XII). Thus it appears that oximes can exist in forms (I) and (II). Brady (1916) considered that oximes in solution are a tautomeric mixture of (I) and (II) (oximino-nitrone diced system). Ultraviolet absorption spectra studies show that the spectra of the oximes are the same as those of the 0-methyl ethers, whereas those of the N-methyl ethers are entirely different. Hence, if oximes are tautomeric mixtures of (I) and (II), the equilibrium must lie almost completely on the oxime side, *i.e.*,

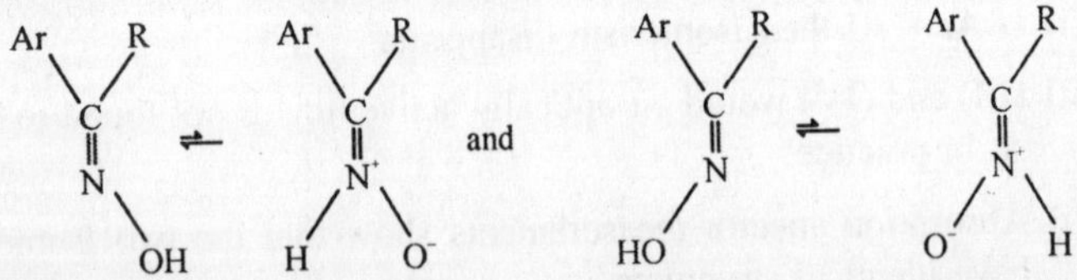

It is possible, however, that none of the nitrone form is present, but its methyl derivative is formed during the process of methylation. If we assume that methyl sulphate provides methyl carbonium ions, then it is possible that these ions attack the nitrogen atom (with its lone-pair) or the oxygen atom (with its two lone-pairs). This would result in the formation of the N- and O-methyl ethers, without having to postulate the existence of the oximino-nitrone tautomeric system.

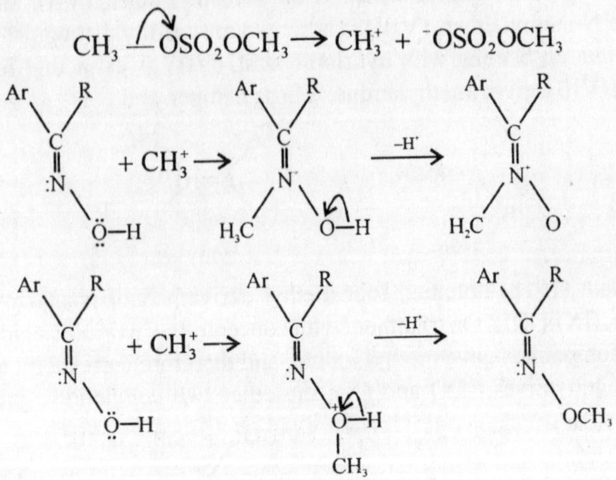

Fig. 2

In terms of modem valency theory, both the carbon and nitrogen atoms in oximes are sp^2-hybridised. This is analogous to the hybridisation of the carbon atoms in ethylene. Hence, the C=N double bond consists of one a- and one n-bond, and the third sp^2 orbital of the nitrogen atom is occupied by a lone pair of electrons. Thus, the oxime molecule is coplanar about the double bond and can exhibit geometrical isomerism (diastereoisomerism).

This geometry is in agreement with the facts described above, and further proof for this configuration is obtained from the examination of the oxime of 4-oxocyclohexane-l-carboxylic acid (XIIIa or b). If the N–O bond is not collinear with the C=N double bond), the configuration is (XIIIa), and it will therefore be optically active. If, however, the three nitrogen valencies are coplanar and symmetrically placed, then the configuration will be (XIIIb), and this will not be optically active, since it possesses a plane of symmetry. Mills and Bain (1910) prepared this oxime and resolved it; hence, its configuration must be (XIIIa).

The problem of geometrical enantiomerism has already been discussed for ethylenic compounds and Lyle *et al.,* (1959) have shown that oximes of ketones of the type Z^+COZ^- can also exhibit this phenomenon. Thus, these authors obtained the (+)-form of 2,6-diphenyl-l-methylpiperid4-one oxime, (XIV); attempts to isolate the (–)-form failed. The mechanism of the interconversion of oximes has been the subject of much debate. A current theory is that the isomerisation involves a 'lateral shift'. The sp2-hybridised nitrogen atom becomes sp-hybridised in the transition state. In this conversion, the lone pair which originally occupied an sp2-orbital now occupies an in-plane p-orbital. The C=N–OH bond angle is 180° and the C, N, and OH remain in the same plane (in the T.S.).

Nomenclature of the Oximes

In oxime chemistry the terms syn and anti are used instead of the terms cis and trans. When dealing with aldoximes, the syn-form is the one in which both the hydrogen atom and the hydroxyl group are on the same side; when these groups are on opposite sides, the configuration is anti. Thus (I) is syn- and (II) is anti-benzaldoxime. With ketoximes, the prefix indicates the spatial relationship between the first group named

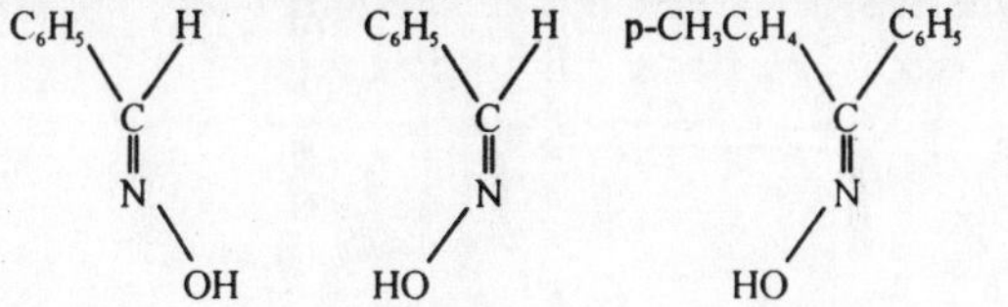

(I) syn (II) anti (III)

and the hydroxyl group. Thus III may be named as syn p-tolyl phenyl ketoxime or anti-phenyl p-tolyl ketoxime.

The E–Z system of nomenclature is also applied to oximes. Thus, the syn-oxime (I) is named benzaldehyde (E)-oxime or (E)-benzaldehyde oxime; (II) is the corresponding (Z)-oxime. The group with the greater priority (phenyl) is taken as being cis with respect to the hydroxyl group. Since p-tolyl has priority over phenyl, (III) is (Z)-p-tolyl phenyl ketoxime.

Determination of the Configuration of Aldoximes

As we have seen, aromatic aldoximes can be obtained in two geometrical isomeric forms, the syn and the anti. Aliphatic aldoximes, however, appear to occur in one form only, and this is, apparently, the anti-form. The problem, then, with aromatic; aldoximes is to assign configurations to the stereoisomeric forms. The two forms (of a given aldoxime) resemble each other in many ways, but differ very much in the behaviour of their acetyl derivatives towards aqueous sodium carbonate. The acetyl derivative of one isomer regenerates the aldoxime; this form is known as the a-isomer.

The other isomer, however, eliminates a molecule of acetic acid to form an aryl cyanide; this form is known as the β-isomer. Hantzsch and Werner (1890) suggested that the β-form readily eliminates acetic acid because the hydrogen atom and the acetoxy-group are close together, *i.e.*, the β-isomer is the syn-form. Such a view, however, is contrary to many experimental results, *i.e.*, the experimental results are:

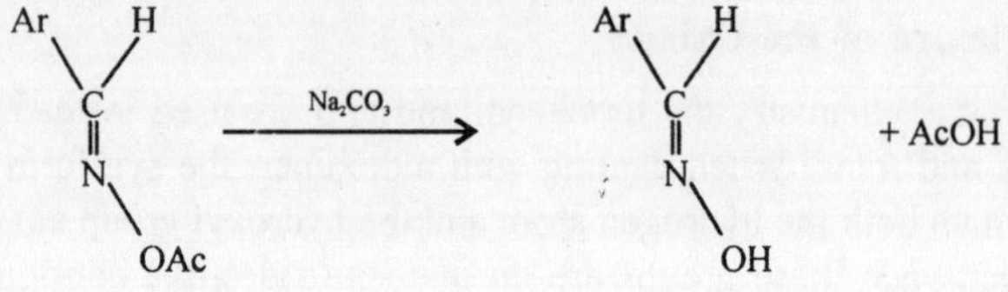

syn- (or E-)

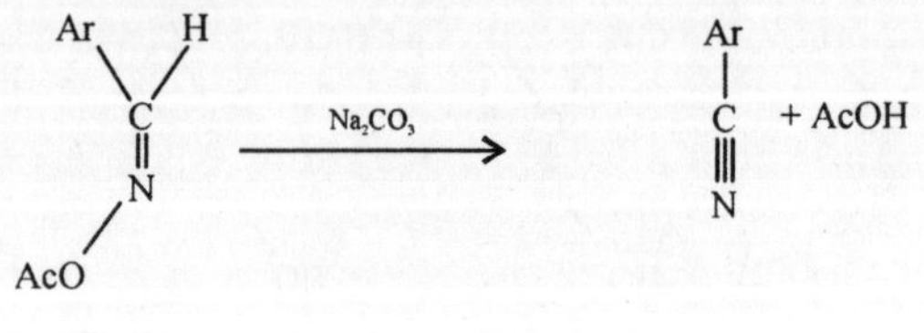

anti - (or Z-)

Fig. 3

Brady and Bishop (1925) found that only one of the two isomers of 2-chloro-5-nitrobenzaldoxime readily gave ring closure on treatment with sodium hydroxide.

It therefore follows that this form is the anti-isomer (cf. method of cyclisation). It was also found that it was this isomer that gave the cyanide on treatment with acetic anhydride followed by aqueous sodium carbonate. Thus antielimination must have occured, *i.e.,* the β-isomer is the anti-form. Actually, the ring compound produced, the 5-nitrobenzisoxazole, is unstable, and rearranges to nitrosalicylonitrile.

In a similar manner, Meisenheimer (1932) found that of the two isomeric 2,6-dichloro-3-nitro-benzaldoximes, it was the anti-isomer that gave ring closure, and was also the one that gave the cyanide. Hence, if anti-elimination is used as the criterion for these reactions, the configurations of the syn- and anti-forms can be determined. It might be noted here, in passing, that since the synform was originally believed to form the cyanide, the configurations of the isomers in the literature up to 1925 (*i.e.,* before Brady's work) are the reverse of those accepted now.

Determination of the Configuration of Ketoximes

The configurations of ketoximes have been mainly determined by means of the Beckmann rearrangement (1886). Aromatic ketoximes, *i.e.,* ketoximes containing at least one aromatic group, occur in two forms; aliphatic ketoximes appear to occur in one form only. When treated with reagents such as sulphuric acid, acid chlorides, acid anhydrides, phosphorus pentachloride, etc., ketoximes undergo a molecular rearrangement, resulting in the formation of an acid amide:

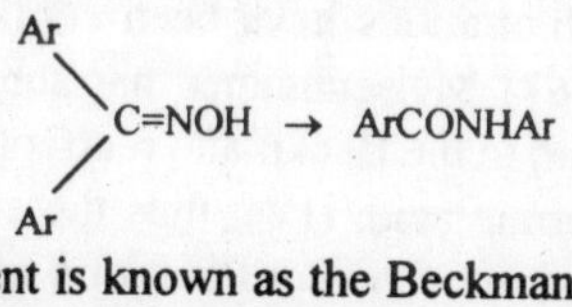

This rearrangement is known as the Beckmann rearrangement. The best method is to treat an ethereal solution of the oxime with phosphorus pentachloride at a temperature below –20°C. On the other hand, Horning *et al.,* (1952) have found that a very good method for effecting the Beckmann rearrangement is to heat the oxime in polyphosphoric acid at 95° to 130°C, and more recently, van Es (1965) has shown that refluxing a solution of the ketoxime in formic acid gives the amide in good yield

Hantzsch (1891) suggested that the course of the rearrangement indicated the configuration of the oxime, and assumed that the syn-exchange of groups occurred since they were closer together in this isomer. This, again, was shown experimentally to be the reverse, *i.e.*, it is the anti-rearrangement that occurs-, and not the syn; thus:

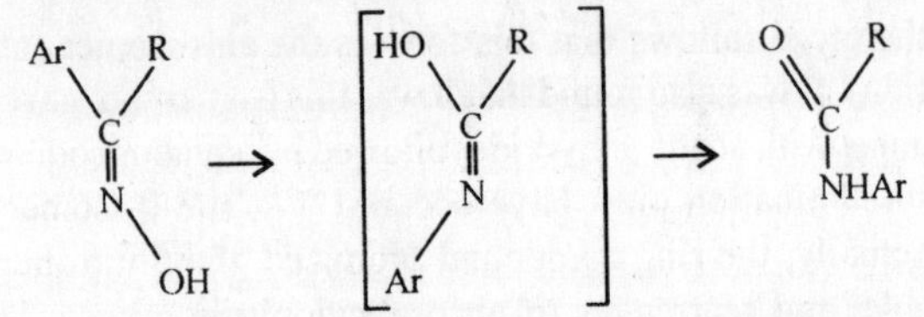

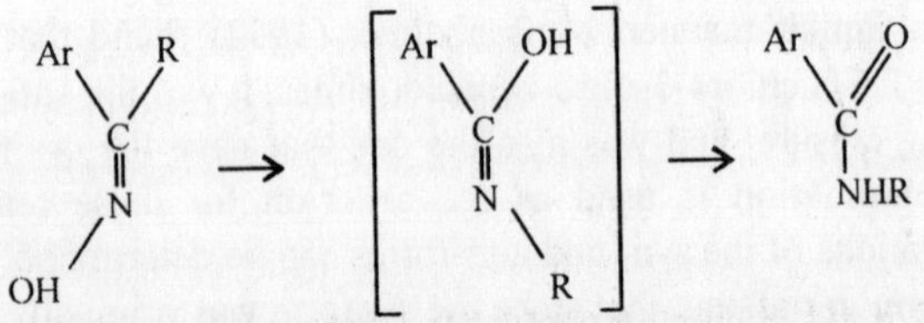

Meisenheimer (1921) subjected triphenylisoxazole, (I), to ozonolysis, and thereby obtained the benzoyl-derivative of anti-phenyl benzil monoxime, (II). This configuration is based on the reasonable assumption that the ozonolysis proceeds without any change in configuration. Furthermore, the monoxime designated the β-isomer gave (II) on

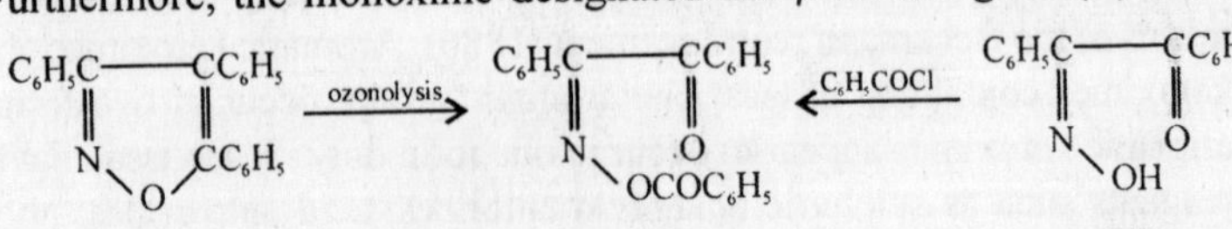

(I) (II) (III)

benzoylation, and so the configuration of the β-isomer, (III), is determined. These assigned configurations have been confirmed by X-ray analysis (Robertson *et al.,* 1967). Meisenheimer then subjected this β-oxime (*i.e.*, the anti-phenyl oxime) to the Beckmann rearrangement, and obtained the anilide of benzoylformic acid, (IV); thus the exchange of groups must occur in the anti-position. The configuration of the β-monoxime, (III), is confirmed by the fact that it may be obtained directly by the

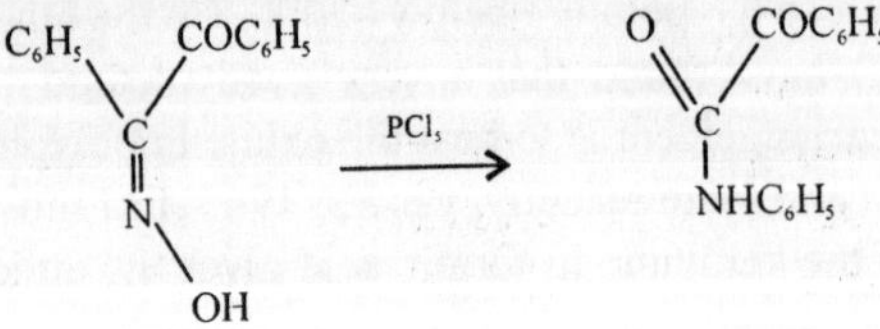

(III) (IV)

ozonolysis of 3,4-diphenyliso-oxazole5-carboxylic acid, (V) (Kohler, 1924). Meisenheimer *et al*., (1925) also demonstrated the antirearrangement as follows.

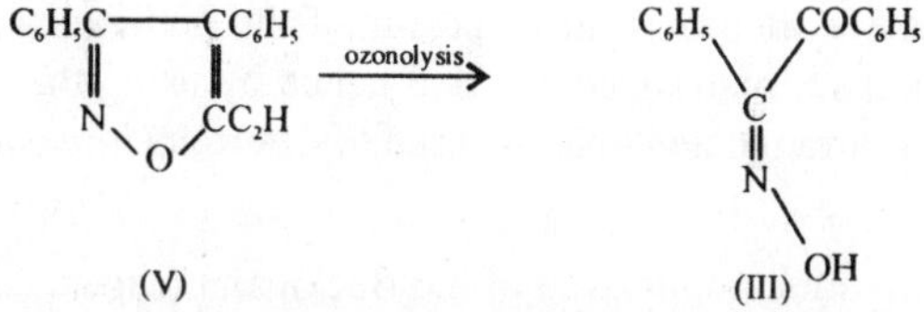

(V) (III)

The α-oxime of 2-bromo-5-nitroacetophenone is unaffected by sodium hydroxide, whereas the fl-isomer undergoes ring closure to form 3-methyl-5-nitrobenziso-oxazole; thus the α-oxime is the syn-methyl isomer (VI) and the α-oxime the anti-methyl isomer (VII). When treated with sulphuric acid or phosphorus pentachloride, the α-oxime underwent the Beckmann rearrangement to give the N-substituted acetamide; thus the exchange occurs in the anti-positions.

Further evidence for the anti-exchange of groups in the Beckmann rearrangement has been obtained by studying the behaviour of compounds exhibiting restricted rotation about a single bond, *e.g,* Meisenheimer *et al.,* (1932) prepared the two isomeric oximes of l-acetyl-2-hydroxy-naphthalene-3-carboxylic acid, (VIII) and (IX), and of these two forms only one was resolvable.

This resolvable isomer must therefore be (IX), since asymmetry due to restricted rotation is possible only with this form. Meisenheimer found that the ethyl ester of (IX), on undergoing the Beckmann rearrangement, gave the amide $ArCONHCH_3$ (where Ar is the naphthalene part of the molecule), whereas the ethyl ester of (VIII) gave the amide $CH_3CONHAr$. These results are in agreement with the anti-exchange of groups in each case. Another method used for examining which groups exchange has been dipole measurements. Sutton *et al.,* (1931) measured the dipole moments of the two isomeric N-methyl ethers of p-nitrobenzophenone oxime and obtained the values shown. These clearly indicate the configurations, and since the oxime corresponding to the nitrone with

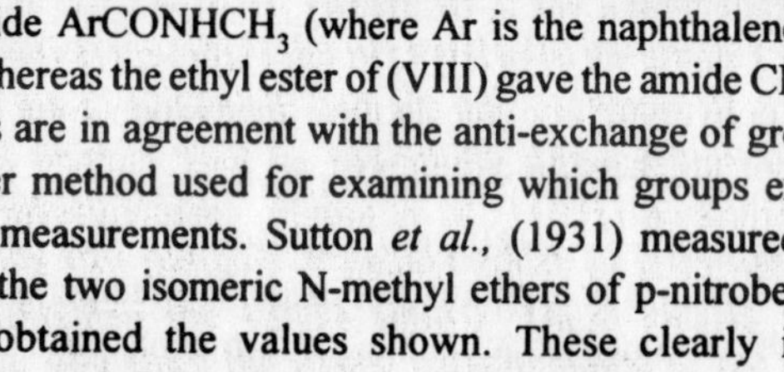

μ=6.60D μ=1.09D

the higher dipole moment gives p-nitrobenzanilide on undergoing the Beckmann rearrangement, and the other isomer gives benzo-p-nitranilide, it follows that the rearrangement occurs by the anti exchange of groups.

Thus the evidence is all in favour of the anti-exchange of groups in the Beckmann rearrangement, and hence by using this principle, the Beckmann rearrangement may be used to determine the configuration of ketoximes.

An interesting application of the Beckmann rearrangement is in the formation of heterocyclic rings, *e.g,* when cyclopentanonoxime is subjected to the Beckmann rearrangement, the nitrogen atom enters the ring (thus producing ring expansion) to form 2-piperidone.

On the other hand, Hill *et al.,* (1956) have shown that the oximes of some spiro-ketones undergo abnormal Beckmann rearrangements in the presence of polyphosphoric acid, *e.g,* spiro-[4, 4]nonanone-1-oxime gives hydrind-8, 9-en-4-one: Although aliphatic ketoximes are not known in two isomeric forms, some may produce two products when subjected to the Beckmann rearrangement, *e.g,* the oxime of pentan-2-one gives N-propylacetamide and N-methylbutyramide. The reason for this is uncertain; possibly oximes of this type are actually a mixture of the two forms; or alternatively, they exist in one stable form which, during the Beckmann rearrangement,

$$\begin{matrix} H_3C \\ CH_3CH_2H_2C \end{matrix}\!\!> C{=}NOH \xrightarrow{PCl_5} CH_3CONHCH_2CH_2CH_3 + CH_3CH_2CH_2CONHCH_3$$

is partially converted into the labile form which then undergoes the rearrangement (cf. benzaldoxime, below).

In an attempt to prepare quinoline by the dehydration of cinnamaldoxime with phosphorus pentoxide, Bamberger and Goldschmidt (1894) actually obtained isoquinoline; the formation of the latter compound and not the former can only be reasonably explained on the assumption that the oxime first undergoes the Beckmann rearrangement, and the rearranged product then undergoes ring closure to form isoquinoline. Later, Horning *et al.,* (1952) have shown that aldoximes can be made to undergo the Beckmann rearrangement under the influence of polyphosphoric acid, *e.g,* syn-benzaldoxime gives a mixture of formanilide and benzamide, the latter being produced by the conversion

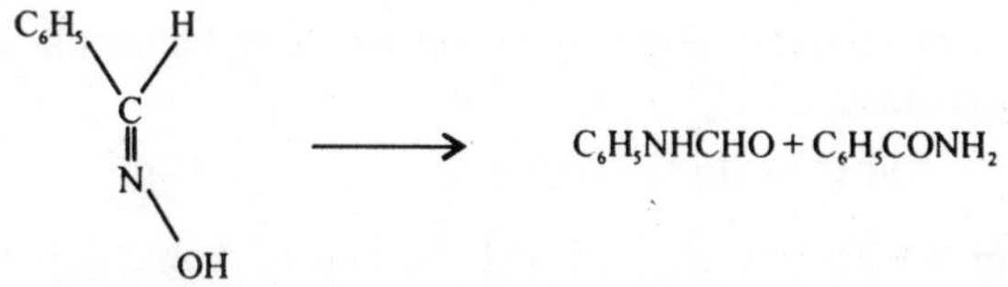

syn- (or E-) isomer

anti- (or Z-) isomer

of the syn-form into the anti; anti-benzaldoxime gives benzamide only. These results are in agreement with the configurations obtained by other methods.

Mechanism of the Beckmann Rearrangement

This rearrangement is an example of the 1,2-shift in which the migration origin is carbon and the migration terminus is nitrogen. As we have seen above an integral part of the rearrangement is the anti migration of the group. Since the oxime itself does not rearrange, it is reasonable to suppose that some intermediate is formed between the oxime and the reagent used to effect the rearrangement, and it is this intermediate which then rearranges.

Kuhara *et al.*, (1914, 1916) prepared the benzenesulphonate of benzophenone oxime and showed that this readily underwent rearrangement in neutral solvents in the absence of any acid catalyst to give an isomeric compound which, on hydrolysis, gave benzanilide and benzenesulphonic acid; thus:

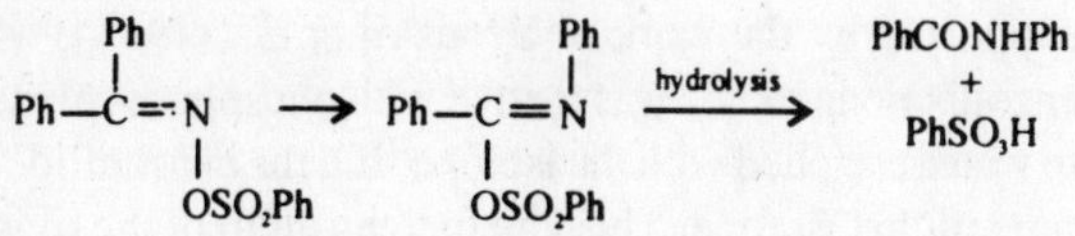

(I)

Kuhara assigned structure (I) to this intermediate on the fact that its absorption spectrum was almost identical with that of the compound prepared by reaction between N-phenylbenzimidoyl chloride and silver benzenesulphonate

$$PhCC1{=}NPh + AgOSO_2Ph \rightarrow (I) + AgC1$$

Kuhara (1926) also showed that the rate of rearrangement of the benzophenone oxime ester is faster the stronger the acid used to form the ester; the order obtained was:

$$PhSO_3H > CH_2ClCO_2H > PhCO_2H > McCO_2H$$

Chapman (1934) showed that the rate of rearrangement of benzophenone oxime picryl ester is faster in polar than in non-polar solvents. Thus the work of Kuhara and Chapman is strong evidence that the rate-determining step in the rearrangement is the ionisation of the intermediate.

Now let us consider the migration of the R or Ar group. This could be either intermolecular or intramolecular, but Kenyon *et al.,* have shown it to be the latter; *e.g,* in 1946, Kenyon *et al.,* showed that when (+)-α-phenylethyl methyl ketoxime is treated with sulphuric acid the product, N-α-phenylethylacetamide, is almost 100 per cent optically pure. Thus the migrating group never separates during the rearrangement, since if it did a racemised product would have been obtained. Furthermore, this retention of optical activity might be cited as evidence for the formation of a bridged-ion during the migration, since in such an ion the migrating group is not free and the 'new partial' bond is formed on the same side as the bond which is breaking (see below).

Another problem that arises here is: Does the anion separate completely during the ionisation or does it also migrate intramolecularly? The work of Kuhara and Chapman strongly suggests complete separation, and this is supported by the work of Brodskii *et al.,* (1941), who found that when benzophenone oxime was treated with phosphorus pentachloride and then with water enriched with the isotope ^{18}O, the benzanilide obtained contained some of this isotope. Thus the oxygen atom of the oxime group must have been completely removed in the ionisation stage. The following mechanism is in agreement with all of the above facts (Y is $PC1_4$, MeCO, etc.); the lower set of equations is the alternative route via a bridged-ion. It might also be noted that when acid is used as the rearranging reagent, OY is possibly OHZ.

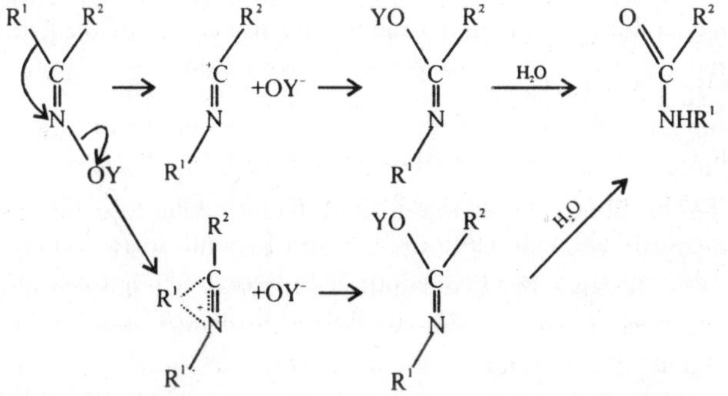

Fig. 4

Support for this mechanism is the evidence obtained for the intermediate formation of the imidoyl ester (RN = CROY); compound (II) was obtained by Heard *et al.*, (1959), who examined the rearrangement of a 17-keto-16-oxime (a steroid). It has been shown that when the migrating group is aryl, the rate of the rearrangement is accelerated when there is an electron-releasing group, *e.g*, Me, in the p-position. This may be cited as evidence to support the formation of a bridged-ion (at least for migrating aryl groups).

Another mechanism has been proposed by Grob *et al.*, (1964) who studied the Beckmann rearrangement of various tosylates. According to these authors, the rearrangement proceeds via a nitrilium salt (Y = OTs). The existence of the nitrilium ion was demonstrated by infrared spectroscopy. This mechanism has been supported by other workers, *e.g*, Schofield *et al.*, (1970) carried out the Beckmann rearrangement of ortho-substituted acetophenone oximes in the presence of sulphuric acid and detected, by infrared spectroscopy, the nitrilium ion as an intermediate.

These workers did not detect the oxime O-sulphonic acid, but this has been done by Yukawa *et al.*, (1971) who used NMR spectroscopy. This mechanism can also be used to explain the course of the rearrangement under the influence of phosphorus pentachloride.

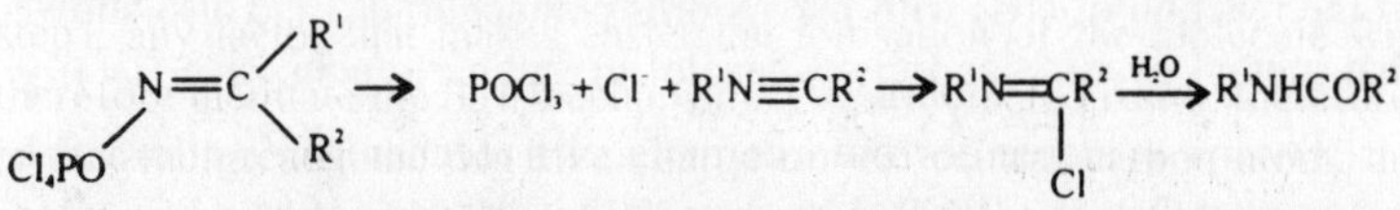

Since phosphoryl chloride can catalyse the Beckmann rearrangement, it is not necessary to use phosphorus pentachloride in molar amount. Actually, Stephen *et al.*, (1956) have shown that one molecule of phosphorus pentachloride, phosphoryl chloride, thionyl chloride, or benzenesulphonyl chloride rearranges two molecules of the ketoxime to yield the corresponding amide and imidoyl chloride in approximately equimolecular amounts, *e.g*,

$$2R_2C{=}NOH + PCl_2 \rightarrow RCONHR + RCCl{=}NR + POCl_3 + HCl.$$

It has also been shown that hydrogen chloride is essential during the rearrangement, but that it does not itself cause the rearrangement of the oxime. On the basis of these results, Stephen *et al.*, have proposed the following mechanism for the Beckmann rearrangement of ketoximes. The reagent first produces some acid amide and imidoyl chloride, and the latter then dehydrates unchanged ketoxime to the anhydride which then reacts as shown: It was also suggested that other reagents which effect the Beckmann rearrangement may function as dehydrating agents for the formation of the ketoxime anhydride.

When a trace of the reagent is used, a large yield of amide is obtained. The mechanism is believed to be the same as that given above, provided that in the initial stage there is sufficient to form a trace of the ketoxime anhydride in the presence of hydrogen chloride. Rearrangement of the anhydride will now take place as above with the formation of the imidoyl chloride which can then dehydrate ketoxime to anhydride, itself being converted into the amide:

$$2R_2C{=}NOH + RCCl{=}NR \rightarrow (R_2C{=}N-)_2O + RCONHR + HCl$$

Thus the yield of amide increases at the expense of the imidoyl chloride.

It can be seen from the foregoing account that several mechanisms appear possible for the Beckmann rearrangement. These are intramolecular, but now an intermolecular mechanism has also been proposed by Hill *et al.*, (1962) who have reported an example in which the migrating group had the inverted configuration in the amide. These authors examined the rearrangement of 9-acetylcis-decalin oxime and have suggested the following mechanism The authors identified methyl cyanide as a product of the reaction of (III) with phosphorus pentachloride, and also showed that methyl cyanide and cis-β-decalol in sulphuric acid gave (IV).

Further evidence for the intermolecular mechanism is afforded by the work of Conley (1963), who showed that cleavage of an oxime which is completely substituted at the α-carbon atom appears to be a general process. When a mixture of phenyl 2-phenylisopropyl ketoxime and pinacone oximc are heated in polyphosphoric acid, the product is a mixture of four secondary amides. Each oxime, on rearrangement under the same conditions, gives only one secondary amide. The formation of the crossed-product from the mixture of oximes indicates a fragmentation-recombination mechanism.

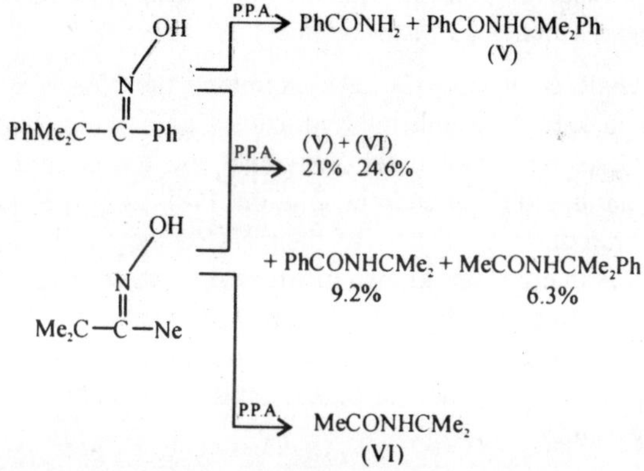

Fig. 5

An unusual example of the Beckmann rearrangement is the action of P.P.A. on the oxime of 4bromo-7-t-butylindan-l-one; the products werę (VII) (predominant product), (VIII) (alkyl migration), and (IX) (aryl migration), with (VIII) being formed to a much greater extent than (IX) (Lansbury *et al.,* 1964).

On the other hand, Lansbury *et al.,* (1966) found that the oxime of 8-t-butyl5-bromotetral-1-one underwent ring expansion (aryl migration) with the loss of the t-butyl group. An interesting reaction associated with ketoximes is the Neber rearrangement (1926). When a ketoxime O-sulphonate is treated with base and then followed by hydrolysis with acid, rearrangement occurs to give an a-amindketone. It appears that the configuration of the oxime has no significant bearing on this reaction.

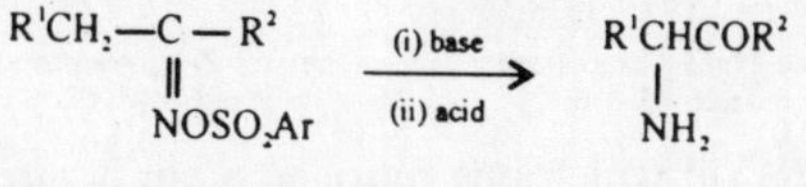

Stereoisomerism of Some Other Tervalent Nitrogen Çompounds Containing a Double Bond

There are several other types of compounds besides the oximes in which the nitrogen atom is linked by a double bond.: he other atom joined by this double bond may be a carbon atom (as in the oximes), or another nitrogen atom, and in both cases stereoisomerism is possible; *e.g*, Krause (1890) obtained two isomeric forms of the phenylhydrazone of o-nitrophenylglyoxylic acid, (I), and Hopper (1925) isolated two isomers of the monosemicarbazone of benzil, (II). Mills and Bain (1914) resolved (III); this is resolvable because of the non-planar configuration of the three nitrogen valencies (cf. the oximes).

Karabatsos *et al.*, (1962) have examined the NM R spectra of a number of ketone dinitrophenyl-hydrazones and semicarbazones, and have distinguished between the syn- and anti-forms, and have also calculated the amounts of each in solution. Phillips (1958) had already examined aldoximes by means of their NMR spectra. Many cases of geometrical isomerism are known in which the two forms are due to the presence of a nitrogennitrogen double bond. Examples of this type which have been most extensively studied are the diazoates, (IV), the diazosulphonates, (V), and the diazocyanides, (VI) for an account of these compounds).

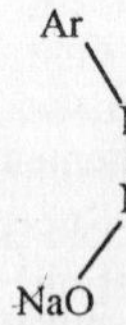

(IV)
syn- (or Z-) form

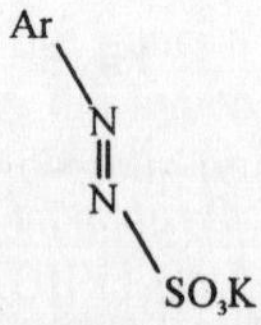

(V)
anti- (or E-) form

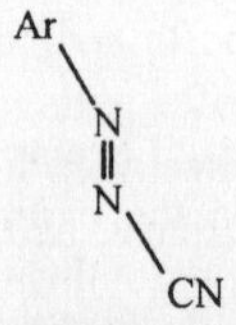

(VI)
anti- (or E-) form

Azobenzene is also an example of this type, and according to Hartley (1938), 'ordinary' azobenzene is the anti-form.

syn- (or Z-) azobenzene
m.p. 71.4°C

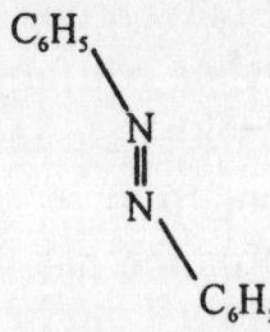

anti (or Z-) azobenzene
m.p. 68°C

Azoxybenzene (in which one nitrogen atom is tercovalent and the other quadricovalent) also exists in two geom:etrical isomeric forms, the anti-isomer being 'ordinary' azoxybenzene.

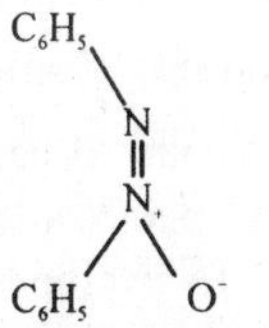

syn- (or E-) azobenzene
m.p. 86°C

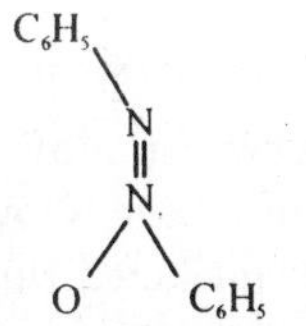

anti- (or Z-) azoxybenzene
m.p. 36°C

Aliphatic nitroso compounds are dimers in the solid state, and Chilton *et al.*, (1955) and Gowenlock *et al.*, (1955) have examined the ultraviolet absorption spectra of the two solid forms of nitrosomethane

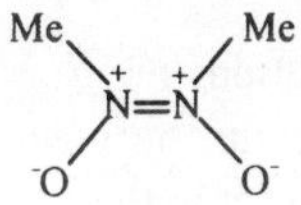

cis (or Z)

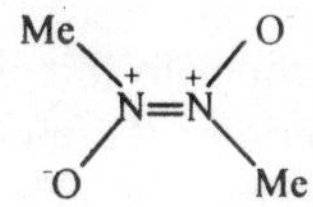

trans (or E)

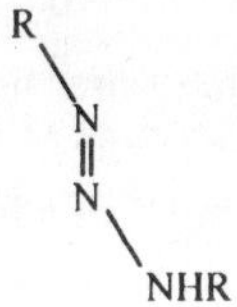

(VII)

and conclude that they are geometrical isomers. This has been confirmed by Liittke (1956, 1957) from his infrared spectra studies.

Recently, Le Fevre *et al.*, (1951) have measured the dipole moments and the ultraviolet absorption spectra of a number of triazens, and have concluded that these compounds exist in the anti-configuration about the nitrogen-nitrogen double bond, *i.e.*, the configuration is (VII). These authors also believe that this anti-form is converted into an equilibrium mixture of the anti- and syn-forms when exposed to sunlight. Harley-Mason *et al.*, (1961) have offered evidence to show that they have isolated the three theoretically possible geometrical isomers of o-nitroacetophenone azine (Ar = o-$NO_2C_6H_4^-$):

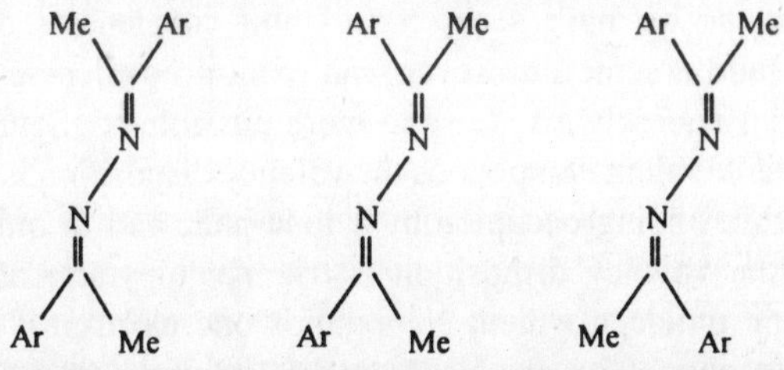

Their evidence was based on infrared, ultraviolet and NMR spectra. This compound appears to be the first example of the isolation and characterisation of all three possible geometrical isomers of an azine.

Conformational Analysis of Ring Systems Containing Nitrogen

The stereochemistry of aziridines has already been discussed. Pyrrolidine (I) has been shown, from spectroscopic data and X-ray analysis, to be a puckered ring similar to cyclopentane and also, like cyclopentane, undergoes pseudorotation. Piperidine (II) has been shown to exist in the chair form (cf. cyclohexane) and was believed to be predominantly in conformation (IIb), *i.e.,* with the hydrogen of the NH group in the axial position. One explanation offered is that the spatial requirements of a lone pair of electrons are greater than those of a hydrogen atom.

There are, however, objections to this explanation, *e.g,* it is to be expected that in N-methylpiperidine the methyl group, because of the 'large bulk' of a lone pair of electrons, would be present in the equatorial form in smaller amount than the corresponding methyl group in methylcyclohexane. In practice, by the use of dipole measurements, it has been shown that hydrogen is slightly more in the equatorial position (IIa) than in the axial (IIb), and methyl is predominantly in the equatorial position. It therefore appears that in heterocyclic ring systems the effect of the lone pair of electrons may be ignored. In certain circumstances, however, the effect of the lone pair must be taken into consideration. This is the case when there is a polar substituent on an adjacent carbon atom.

As we have seen because of intramolecular hydrogen bonding, cyclohexane-1,4-diol is in the boat form (or twist-boat). In some highly substituted 4-hydroxypiperidines the boat form is also present, *e.g,* (III).

STEREOCHEMISTRY OF PHOSPHORUS COMPOUNDS

Nitrogen, as we have seen, can exhibit covalencies of 3 and 4; phosphorus (and arsenic), however, can exhibit covalencies of 3, 4, 5 and 6, and consequently gives rise to more possible configurations than nitrogen. In tercovalent compounds the valency disposition is tetrahedral (sp^3), one orbital being occupied by a lone-pair; and in quinquevalent compounds the valency disposition is trigonal bipyramidal (sp^3d). In quadricovalent unielectrovalent compounds one electron is transferred from the phosphorus or arsenic atom to the anion and the valency disposition is tetrahedral (sp^3). When there are double bonds present, one is a σ- and the other is apn-dr-bond; thus, in $POCl_3$, the shape is tetrahedral.

Tercovalent Phosphorus Compounds

Since the electronic configuration of phosphorus is $(1s^2)(2s^2)(2p^6)(3s^2)(3p^3)$, it might be expected that suitable tercovalent compounds, R_3P, could be resolved, since the configuration would be a trigonal pyramid. However, the phosphorus atom is in a state of oscillation. Calculation has shown that the frequency of this oscillation in phosphine is 5×10^{10}; this is slower than that of nitrogen (2.3×10^{10}), and if it could be brought to zero, then tertiary phosphines would be resolvable. Increasing the weight of the groups slows down the oscillation in phosphorus compounds, *e.g,* replacement of the three hydrogen atoms by deuterium atoms changes the frequency to 6×10^3.

It seems possible, therefore, that very large groups might produce phosphines which would be resolvable. This has been shown to be the case in practice, *e.g,* Horner *et al.,* (1961) resolved EtMePhP, MePhPrP, etc., and (1966) obtained the optically active diphosphine, $MePhPCH_2CH_2PPhMe$. Horner was also able to determine the absolute configuration of MePhPrP as follows. When an enantiomer of an optically active phosphine is treated with benzyl bromide, the phosphonium bromide is obtained with retention of configuration. This salt, on electrolytic reduction, regenerates the original phosphine without loss of optical activity. Now, the absolute configuration of (+) benzylmethylphenylpropylphosphonium bromide has been shown to be (S). Hence, the phosphine obtained also has the (S)-configuration.

$$\underset{(S)-(+)}{\text{Me Pr PhP}} \xrightarrow[\text{2e; H}^+]{\text{PhCH}_2\text{Br}} \underset{(S)-(+)}{\text{Me Pr Ph}\overset{+}{\text{P}}\text{CH}_2\text{Ph}\}\text{Br}^-}$$

Optically active phosphines are fairly stable (optically); they are racemised on heating (probably through oscillation).

Quadricovalent and Quinquevalent Phosphorus Compounds

The earliest phosphorus compounds to be resolved were the phosphine oxides, *e.g,* Meisenheimer *et al.,* (1911) resolved ethylmethylphenylphosphine oxide (I) and benzylmethylphenylphosphine oxide (II).

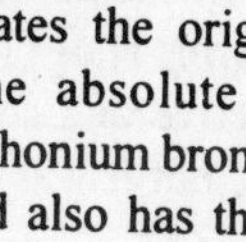

(I)

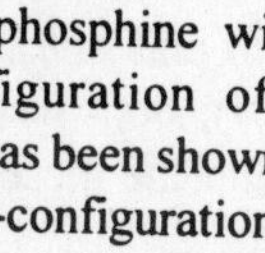

(II)

Many optically active compounds containing quinquevalent phosphorus of the type $R_3P{=}Z$ have now been prepared, *e.g,*

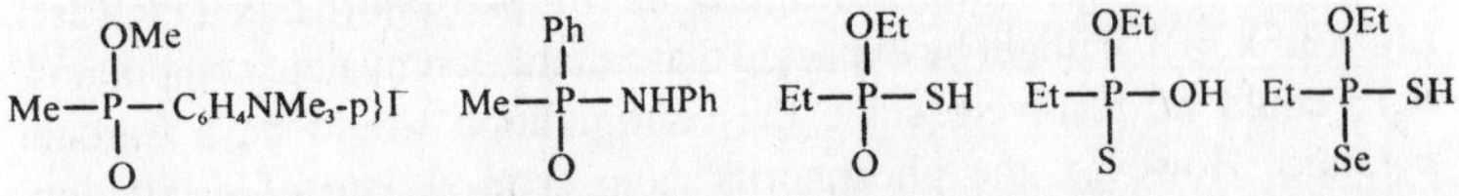

Optically active phosphine oxides are racemised by acids. A possible mechanism is via the formation of the symmetrical compound which has a trigonal bipyramid configuration sp^3d. This is supported by the fact that when the aqueous solution contained $H_2\,^{18}O$, ^{18}O was incorporated in the product (Denney *et al.,* 1964).

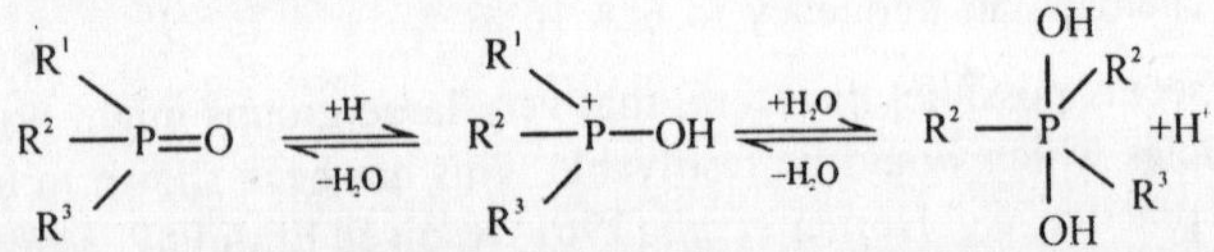

Another interesting phosphorus compound from the point of view of optical isomerism is ethyl triphenylmethylpyrophosphonate (III). If the two phosphorus atoms are asymmetric, then (III) contains two similar asymmetric carbon atoms, and so its structure corresponds to the molecule *CabdCabd.* Thus there will be one racemic modification (composed of the pair of enantiomers) and one mesoform. Hatt (1933) obtained two forms of compound (III); both were inactive and so correspond to the racemic modification and the meso-form, but it was not possible to tell which was which.

Many attempts have been made to resolve quaternary phosphonium compounds, but until recently, all these attempts failed. This failure is attributed to the occurrence in solution of a 'dissociation-equilibrium', which causes very rapid racemisation.

$$[abcdP]^+\,X^- \rightleftharpoons abcP + dX$$

The earlier attempts to resolve phosphonium compounds were always carried out on compounds containing at least one alkyl group; consequently dissociation in solution could occur, thereby resulting in racemisation. Holliman and Mann (1947) overcame this difficulty by preparing a much more stable type of phosphonium compound; these workers prepared a salt in which the phosphorus atom was in a ring, viz., 2 p-hydroxyphenyl-2-phenyl-1,2,3,4-tetrahydro-isophosphinolinium bromide, (IV), and resolved it. The resolution of 4-covalent compounds of phosphorus does not prove that the phosphorus atom has a tetrahedral configuration; it only proves that the phosphorus atom cannot be in the same plane as the other four groups attached to it. Mann *et al.,* (1955), however have now synthesised P-spirobis-1,2,3,4 tetrahydro-phosphinolinium iodide (V) and resolved it into (+)- and (–)-forms which have high optical stability. The phosphorus atom is not asymmetric in this compounds; it is the tetrahedral disposition of the four valencies which produces the dissymmetric cation.

It is interesting to note, in view of what has been said above about the dissociation of phosphonium salts containing at least one alkyl group,

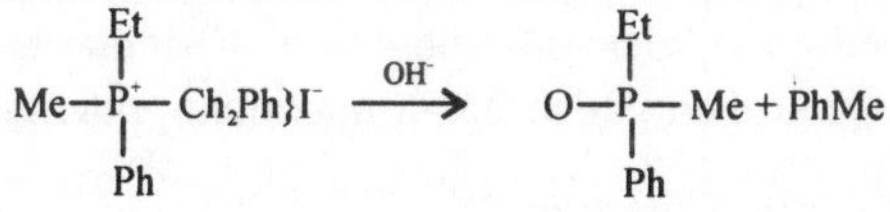

(VI) (VII)

that McEwen *et al.,* (1959) have resolved benzylethylmethylphenylphosphonium iodide (VI). Also, McEwen *et al.,* (1964) have shown that this compound, on treatment with sodium hydroxide, undergoes inversion to give ethylmethylphenylphosphine oxide (VII) and toluene. Homer *et al.,* (1965) have prepared (VIII) in its optically active forms and showed, by X-ray analysis, that the (+)-enantiomer has the (S)-configuration. Phosphines are readily oxidised to phosphine oxides by hydrogen peroxide with retention of configuration. Hence it is possible to prepare an enantiomer of a phosphine oxide with a known absolute configuration, *e.g,*

$$\underset{(S)\text{-}(+)}{\text{Me Pr PhP}} \xrightarrow{H_2O_2} \underset{(S)\text{-}(+)}{\text{Me Pr PhP} = \text{O}}$$

Since phosphines combine directly with sulphur to form phosphine sulphides (with retention of configuration), the absolute configurations of these sulphides may also be determined.

Campbell *et al.,* (1960) have prepared a series of azophosphaphenanthrens ((IX); *e.g,* $R^1 = H$, $R^2 = NMe_2$), but could not resolve them. When the phosphine (IX) was oxidised with hydrogen peroxide, the phosphine oxide obtained, (X), was resolved. Reduction of the (+)-oxide with lithium aluminium hydride gave the (–)-phosphine (IX), and in the same way the reduction of the (–)-oxide gave the (+)-phosphine (IX). It is not certain whether the optical activity in (IX) is due to an asymmetric tervalent phosphorus atom or to a rigid puckering of the molecular framework, which is a 2,2'-bridged biphenyl. A tercovalent phosphorus compound which does not exhibit optical isomerism, but exists as geometrical isomers is 5,10-diethyl-5, 10-dihydrophosphanthren, (XI). It is folded about the P–P axis, and Mann *et al.*, (1962) isolated two forms (see the corresponding arsenic compounds, (X)–(XII), for further discussion). Hellwinkel (1965) has resolved salts of (XII). The anion contains hexavalent phosphorus and has an octahedral configuration.

STEREOCHEMISTRY OF ARSENIC COMPOUNDS

Arsenic, like phosphorus, can exhibit covalencies of 3, 4, 5 and 6; consequently these two elements show a great similarity to each other, and differ from nitrogen which has a maximum valency of 4.

Tercovalent Arsenic Compounds

The electronic configuration of arsenic is $(1s^2)(2s^2)(2p^6)$-$(3s^2)(3p^6)(3d^{10})(4s^2)(4p^3)$. Thus the configuration of tercovalent arsenic compounds will be a trigonal pyramid. Physico-chemical evidence (X-ray analysis, spectroscopy and electron diffraction) has shown that in tercovalent compounds the arsenic atom is at the apex of a tetrahedron, and that the intervalency angle is $100 \pm 4°$. It has also been shown that the arsenic is in a state of oscillation, the frequency of this oscillation through the plane of the three hydrogen atoms in arsine being 16×10^4. This is slower than that of phosphorus (5×10^6), and very much slower than that of nitrogen (2.3×10^{10}).

Thus, preventing the oscillation of the arsenic atom, possibly by attachment to very large groups, should lead to the isolation of optically active tercovalent compounds. In fact, calculations by Weston (1954) led him to the conclusion that tervalent arsenic (and antimony and sulphur) compounds should be stable to inversion at room temperature. Horner *et al.*, (1962) have prepared the optically active arsines EtMePhAs and n-BuMePhAs by removal of the benzyl group from the corresponding optically active arsonium compound and Mislow *et al.*, (1963) have resolved (I). Chan (1968) has also prepared optically active t-BuMePhAs t'la a chiral platinum (II) complex.

Heterocyclic 9,10-dihydroanthracenes have been resolved. Thus, Lesslie *et al.*, (1934) resolved 10-methylphenoxarsine-2-carboxylic acid, (II). These authors suggested that the asymmetry of the molecule is due to a 'butterfly' configuration, *i.e.*, the molecule is folded about the O—As axis (see IV). The authors also pointed out that there was the possibility that the asymmetry might be due to the presence of a stable 'asymmetric' arsenic atom, but preferred the former explanation. Molecules of the type

(II) cannot be coplanar unless the oxygen and arsenic valency angles are both 120°. Since this is not the case (∠COC is approximately 104°), the butterfly configuration is reasonable. There are, however, some difficulties, *e.g,* the corresponding thianthren compounds, phenoxathiin (replace As by S) and phenoxaselinin (replace As by Se) cannot be resolved. Dipole moment measurements of these compounds showed that they have a folded structure and consequently the failure to resolve them was ascribed to instability of the folded structure, this readily undergoing rapid racemisation by `fluttering' through the planar conformation. Since (II) and also [(III); Mislow *et al.,* 1963) have been resolved, it might appear that in such arsenic compounds the folded conformation is stable.

This could be supported by the argument that optically active acyclic arsines have now been prepared (and also the isolation of the two forms (X) and (XI); see below). Mislow, however, has presented evidence that the folded conformation of phenoxarsines is highly flexible, and Campbell (1968) has pointed out that if the folded conformation were stable, then two racemates are possible. Since only one has been isolated, this supports the presence of a stable 'asym metric' arsenic atom in an unstably folded molecule in which the butterfly wings can flutter.

According to Mislow, resolvable phenoxarsines are most correctly pictured as separable configurationally stable enantiomers which individually exist in solution as mixtures of rapidly interconverting diastereoisomeric folded conformations (the term configuration refers to the arsenic pyramid). When each enantiomer of (IV) is treated with ethyl iodide, the same racemised product is obtained. This is due to the fact that when the arsonium compound, (V), is formed, the asymmetric quaternary arsenic atom is racemised owing to the dissociation-equilibrium. Lesslie and Turner (1936) also resolved 10-phenylphenoxarsine-2-carboxylic acid, (VI). This compound was very stable, and oxidation to the arsine oxide, (VII), gave a completely racemised product.

Campbell *et al.,* (1956) have resolved some substituted 9-arsafluorenes, *e.g,* 9 p-carboxyphenyl-2methoxy-9-arsafluorene (VIII). Campbell (1956) has also resolved 2-p-carboxyphenyl-5-methyl1,3-dithia-2-arsaindane (IX). This compound is optically stable in chloroform solution, but is racemised in aqueous sodium hydroxide. Campbell believed that this racemisation is due to the fission of the As-S bonds by aqueous alkali, and subsequent reversal of the reaction by acid, a type of behaviour observed in triaryl thioarsenites (Klement *et al.,* 1938). Furthermore,

Cohen *et al.*, (1931) have shown that in sodium hydroxide solution, alkyl thioarsenites exist in equilibrium with thiol and arsenoxide:

$$R^1As(SR^2)_2 + 2H_2O \underset{H^+}{\overset{OH^-}{\rightleftharpoons}} R^1As(OH)_2 + 2R^2SH$$

Chatt and Mann (1940) prepared 5,10-di-p-tolyl-5,10-dihydroarsanthren, and pointed out that if the valency angle of arsenic remains constant at its normal angle (of approximately 100°), then the structure will be folded, and consequently the three geometrical isomers, (X), (XI) and (XII), are apparently possible (T represents the p-tolyl group). Chatt and Mann also pointed out that evidence obtained from models constructed to scale showed that the two p-tolyl groups (T) in (XII) would almost be coincident, and hence this isomer cannot exist. These authors isolated two optically inactive forms, but were unable to say which was which. When each compound was treated with bromine, both gave the same tetrabromide which, on hydrolysis, gave only one tetrahydroxide. The loss of isomerism in the tetrabromide (and in the tetrahydroxide) may be explained as follows. Bromination of (X) and (XI) converts tercovalent arsenic into quinquecovalent arsenic, and in the latter state the ring valency angles of the arsenic become 120°, and so the arsanthren nucleus is now planar.

Thus both the forms (X) and (XI) would give the same tetrabromide (XIII) (the same is true for the tetrahydroxide); the tetrabromide should thus be planar, the configuration of each arsenic atom being trigonal bipyramidal in the 5-covalent state. Quinquevalent phosphorus and arsenic can make use of the 3d or 4d orbitals, respectively. Thus nitrogen has a maximum covalency of 4, whereas that of phosphorus and arsenic is 5 or 6, *e.g*, the covalency of 6 is exhibited by phosphorus in solid phosphorus pentachloride; X-ray diffraction shows this `molecule' (in the solid state) is $PCl_4^+PCl_6^-$.

Phosphorus, which is $(ls^2)(2s^2)(2p^6)(3s^2)(3p^3)$ in the ground state, may become $(1s^2)(2s^2)(2p^6)(3s)(3p^3)(3d)$ in its 'valence state', since the 3s and 3d orbitals have energy levels which are close together. Kimball (1940) showed, by calculation, that this arrangement, *i.e.*, sp^3d, could give rise to the stable trigonal bipyramidal configuration. This consists of three equivalent coplanar orbitals pointing towards the corners of an equilateral triangle, and two orbitals perpendicular to this plane. Electron diffraction studies of the vapours of phosphorus pentachloride and pentafluoride indicate the trigonal bipyramidal configuration in these molecules. The phosphonium ion might possibly be formed from this trigonal bipyramid by the transference of one of the electrons, or by the transference of a 3s electron and hybridisation of the (3s) ($3p^3$) orbitals; in either case the tetrahedral configuration of the phosphonium ion can be asymmetric, but only in the case of the hybridisation of the (3s)($3p^3$) orbitals will the four bonds be equivalent. Since the properties of phosphonium compounds are in agreement with the equivalence of the four bonds, it therefore appears, on theoretical grounds, that the tetrahedral configuration with the phosphorus atom at the centre is the probable one.

From the experimental side, the preparation of optically active spiro-compounds of phosphorus and of arsenic proves the tetrahedral configuration of these atoms. Earlier work by Mann *et al.*, (1936, 1937) has also definitely established this configuration. These authors prepared compounds of the type $[R_3As\text{-}CuI]_4$ by combination of tertiary arsines or phosphines with cuprous iodide (or silver iodide); in these compounds the phosphorus or arsenic is 4-covalent, and X-ray analysis studies of the arsenic compound showed that the arsenic atom is at the centre of a tetrahedron. Since the corresponding phosphorus compounds are isomorphous, the configuration of the phosphorus is also tetrahedral.

Horner *et al.*, (1965) have been able to assign absolute configurations to benzylmethylphenylpropylarsonium salts by means of the quasi-racemate method (with the corresponding phosphonium compounds; Formula (VIII).

In the solid state, phosphorus and arsenic compounds may contain a negatively charged phosphorus or arsenic atom, *e.g,* $PCl_4^+PCl_6^-$. In this condition, the phosphorus acquires an electron to become –(3s)($3p^3$)($3d^2$), and the arsenic also acquires an electron to become–(4s)($4p^3$)($4d^2$). In both cases the configuration is octahedral (six sp^3d^2 bonds), *e.g,* the following compound has been resolved (Rosenheim et al., 1925); see also the-P (vi) compound (XII).

5

Monochronic Optical Rotations

INTRODUCTION

As the methyl group shifts nearer to the centre of the chain, the molecular rotations get smaller and smaller. A consequence of the distance rule is that molecular rotations of higher members of homologous series containing one chiral centre tend to reach a limiting value or zero value, *e.g,* long-chain fatty acids containing an a-methyl group have molecular rotations which approach the value of ~28°. The application of the method of monochromatic optical rotations is based on the above generalisation. This may be restated as follows: If two compounds have the same absolute configuration, then if their structures differ only at some distance from their chiral centres, the molecular rotations have the same sign and have approximately the same magnitude, *e.g,* acyclic secondary alcohols with the formula shown in (I; m > n) are all dextrorotatory (for the sodium D-line).

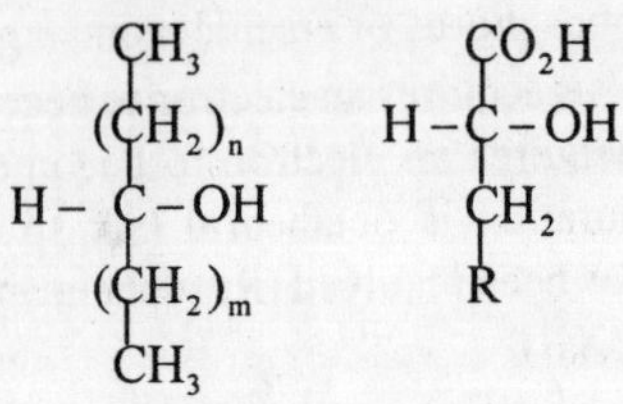

(I) (II)

A very interesting point about this generalisation is that deviations are most likely to occur when the structural difference involves the introduction of a group that absorbs in the near ultraviolet. Thus, the hydroxy-acids represented by (II), in which R = H, Me, Et, Pr, etc., are all laevorotatory, whereas when R = Ph, the acid is dextrorotatory.

SPECIFICATION OF ABSOLUTE CONFIGURATIONS

Since the configuration of (+)-tartaric acid has been related to that of (+)-glyceraldehyde and since the absolute configuration of (+)-tartaric acid has been determined it is now possible to assign absolute configurations to many compounds whose relative configurations to (+)-glyceraldehyde are known. This raised the problem of using one system of specifying absolute configurations. Cahn *et al.*, (1956, 1964) have proposed such a system and this is now widely used. Let us first consider the procedure for a molecule containing one asymmetric carbon atom (one chiral centre).

(i) The four groups are first ordered according to the sequence rule. According to this rule, the groups are arranged in decreasing atomic number of the atoms by which they are bound to the asymmetric carbon atom. If two or more of these atoms have the same atomic number, then the relative priority of the groups is determined by a similar comparison of the atomic numbers of the next atoms in the groups (*i.e.*, the atoms joined to the atom joined to the asymmetric carbon atom). If this fails, then the next atoms of the group are considered.

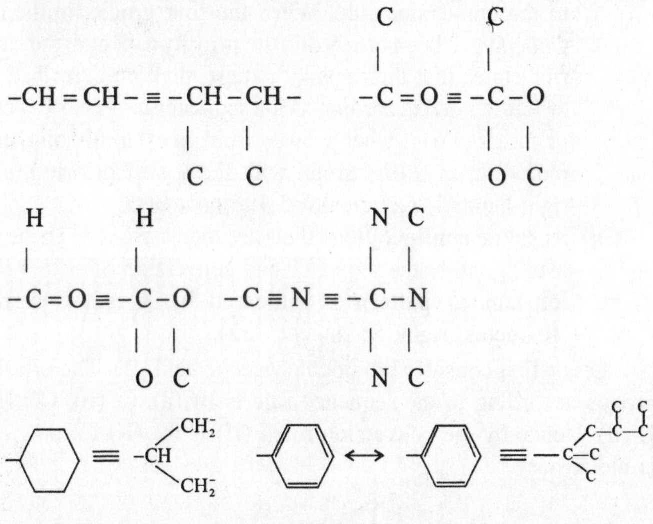

Fig. 1

Thus one works outwards from the asymmetric carbon atom until a selection can be made for the sequence of the groups. When multiple bonds or rings are present, the procedure for determining priority is as follows. Both atoms attached to the multiple bond are considered to be duplicated (for a double bond) or triplicated (for a triple bond), *e.g,*

The priority sequence is then determined by consideration of the duplicated or triplicated 'structure' in which there arephantom-atoms, *e.g,* –CHO is –CH(O)–O(C) and –$CH(OH)_2$ is –CHOH–OH (the phantom atoms in the former are those in parentheses). Both groups contain a carbon atom joined to two oxygen atoms, but since C precedes H, –CHO precedes –$CH(OH)_2$ in priority.

Ring systems are treated as branched chains, and if unsaturated, then duplication is used for a double bond (or triplication for a triple bond). By using these rules, it can be shown that the order of priority sequence (for some of the common substituents) is: I, Br, Cl, SO_3H, SH, F, OCOR, OR, OH, NO_2, NR_2, NHR, NH_2, CO_2R, CO_2H, COR, CHO, CH_2OH, CN, Ph, CR_3, CHR_2, CH_2R, CH_3, D, H.

(ii) Next is determined whether the sequence describes a right- or left-handed pattern on the molecular model as viewed according to the conversion rule. When the four groups in the molecule *Cabcd* have been ordered in the priority a, b, c, d, the conversion rule states that their spatial pattern shall be described as right- or left-handed according as the sequence a → b → c is clockwise or anticlockwise when viewed from an external point on the side remote from d (the group with the lowest priority), *e.g,* (I) in right-handed (*i.e.,* clockwise) arrangement.

(iii) Absolute configuration labels are then assigned. The asymmetry leading under the sequence and conversion rules to a right- and left-handed pattern is indicated by R and S respectively (R; rectus, right; S; sinister, left).

Let us first consider bromochloroacetic acid (II). The priority of the groups according to the sequence rule is Br (a), Cl (b), CO_2H (c) and H (d). Hence by the conversion rule,, (II) is the (R)-form (a → b → c is clockwise).

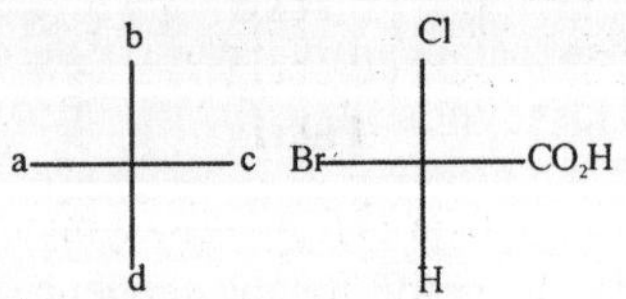

(II)

Now let us consider D(+)-glyceraldehyde. By convention it is drawn as (III) (this is also the absolute configuration). Reference to the sequence list gives the priority sequence : OH (a), CHO (b), CH_2OH (c), and H (d). Since the interchanging of two groups inverts the configuration, the sequence (III) → (IV) → (V) gives the original configuration. Since (V) corresponds to (VI), it thus follows that D(+)-glyceraldehyde is (R)-glyceraldehyde.

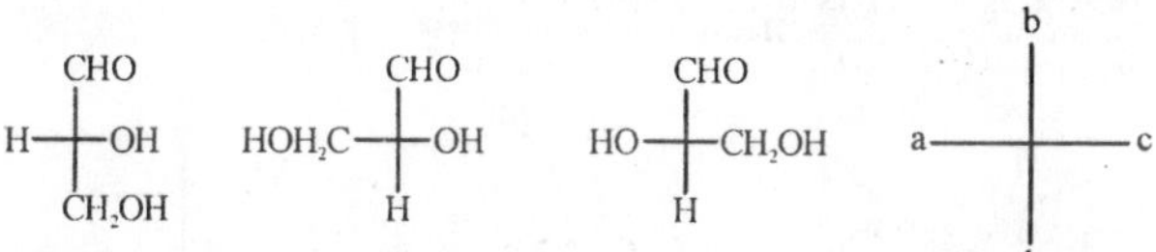

D(+)- (III) L(–)- (IV) D(+)- (V) (VI)

This scheme can be applied to the deutero compound (VII), which is therefore the (R)-form.

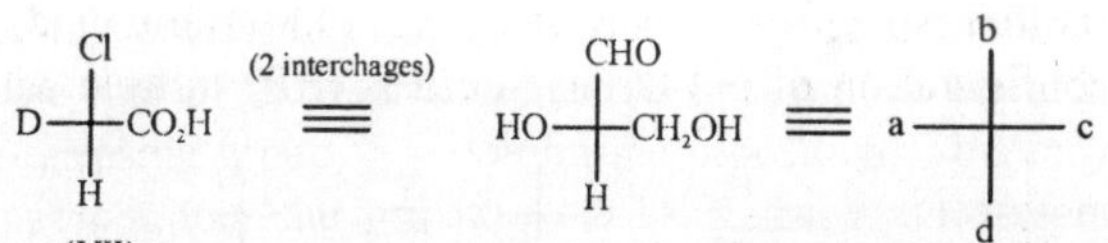

(VII)

On the other hand, (VIII) is the (S)-form, since a → b → c is anticlockwise.

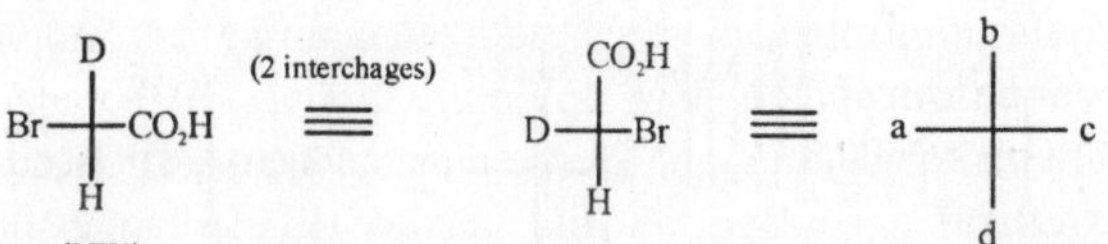

(VIII)

By reference to the sequence list above, it. can be seen that (IX) is the (S)-form.

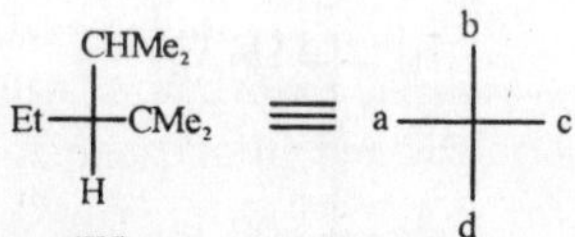

(IX)

Now let us consider some ring systems. As pointed out above, these systems are treated as branched chains, etc. Hence (X) is the (S)-form, since the CHOH group in the left-hand ring is reached before that in the right-hand ring.

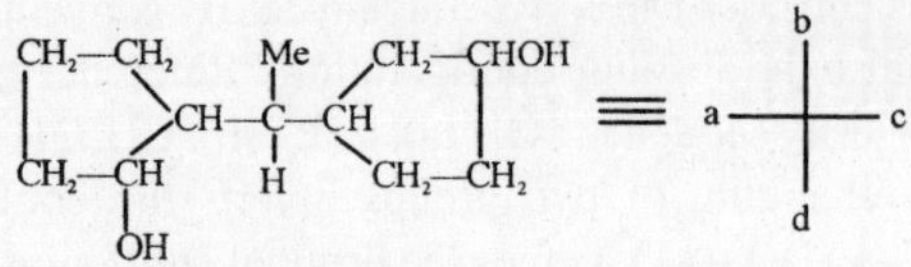

(X)

The same procedure is used when the asymmetric carbon atom is in the ring, *e.g,* (XI) is the (S)-form.

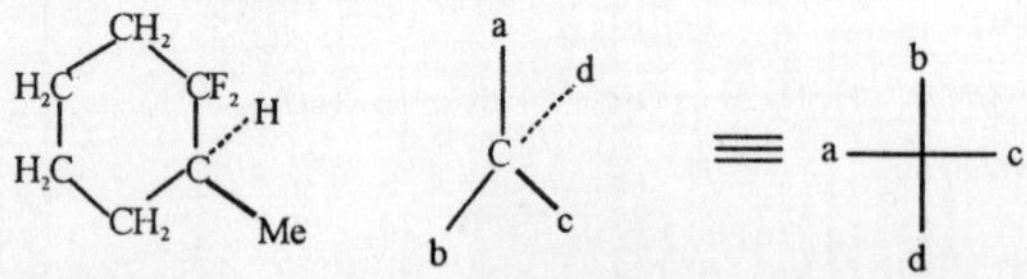

(XI)

When a molecule contains two or more chiral centres, each chiral centre is assigned a configuration according to the sequence and conversion rules and is then specified with R or S, *e.g,* (+)-tartaric acid. Thus the absolute configuration of (+)-tartaric acid is (RR)-tartaric acid:

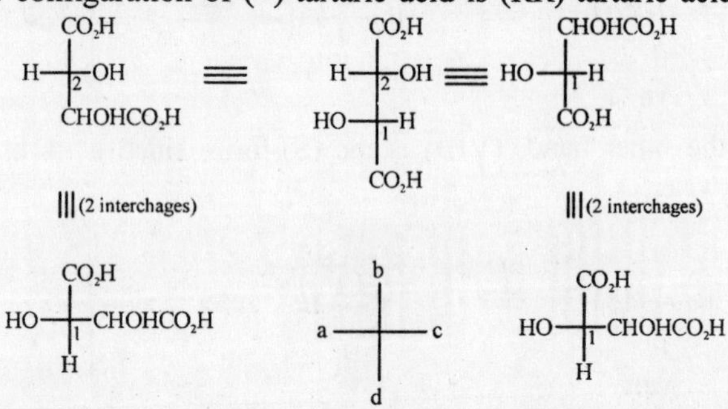

In a similar way it can be demonstrated that n(+)-glucose has the absolute configuration shown.

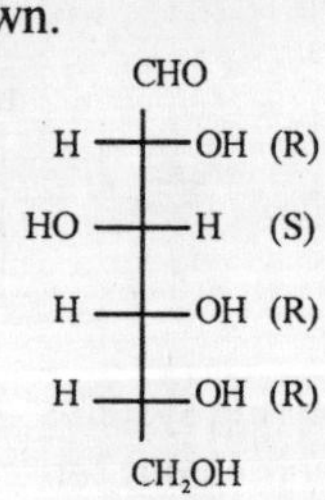

D(+)-gluose

The sequence rule was designed to relate the symbols D and L with the symbols R and S. However, D and L are obtained by means of chemical transformations, whereas R and S are derived from geometrical models and are independent of correlations. Because of this, R and S must he applied only to compounds whose absolute stereochemistry has been determined; they do not necessarily correlate chemical families, *e.g,* (+)-tartaric acid, whether it be n or L (according to the method of correlation) has an absolute configuration specified by (RR). It should be noted that in the same way as n and L are not necessarily connected with the direction of rotation, nor are R and S. The absolute configurations of chiral molecules which do not contain asymmetric carbon atoms may also be specified by an extension of the system described above.

The, test of superimposing a formula (tetrahedral) on its mirror image definitely indicates whether the molecule is symmetric or not; it is asymmetric if the two forms are not superimposable. The most satisfactory way in which superimposability may be ascertained is to build up models of the molecule and its mirror image. Usually this is not convenient, and so, in practice, one determines whether the molecule possesses (i) a plane of symmetry, (ii) a centre of symmetry or (iii) an alternating axis of symmetry. If the molecule contains at least one of these elements of symmetry, the molecule is symmetric; if none of these elements of symmetry is present, the molecule is asymmetric.

It should be remembered that it is the Fischer projection formula that is normally used for inspection. As pointed out is necessary, when dealing with conformations, to ascertain whether at least one of them has one or more elements of symmetry. If such a conformation can be drawn, then the compound is not optically active.

(i) **A plane of symmetry** divides a molecule in such a way that points (atoms or groups of atoms) on the one side of the plane form mirror images of those on the other side. This test may be applied to both solid (tetrahedral) and plane-diagram formulae, *e.g,* the plane-formula of the meso-form of Cabd possesses a plane of symmetry; the other two, (+) and (–), do not.

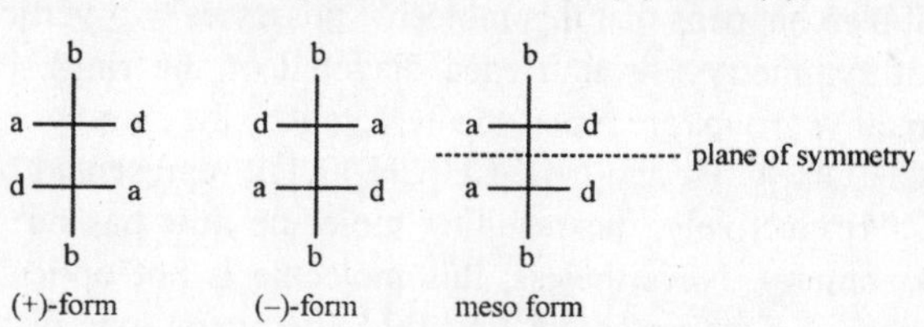

Fig. 2

(ii) **A centre of symmetry** is a point from which lines, when drawn on one side and produced an equal distance on the other side, will meet identical points in the molecule. This test can be satisfactorily applied only to three-dimensional formulae, particularly those offering systems, *e.g,* 2,4-dimethylcyclo-butane-l,3-dicarboxylic acid. The form shown possesses a centre of symmetry which is the centre of the ring. This form is therefore optically inactive.

Another example we shall consider here is dimethyldiketo-piperazine; this molecule can exist in two geometrical isomeric forms, cis and trans. The cis-isomer has no elements of symmetry and can therefore exist in two enantiomeric forms; both are known. The trans-isomer has a centre of symmetry and is therefore optically inactive.

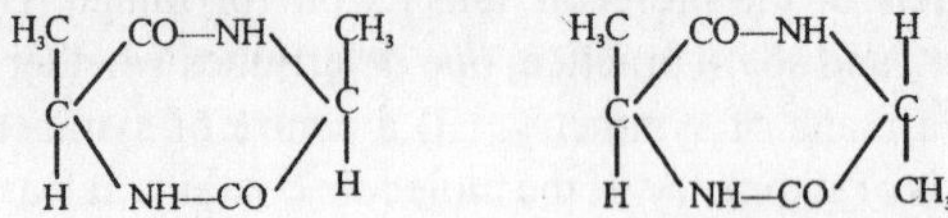

cis trans

It is important to note that only even-membered rings can possibly possess a centre of symmetry.

(iii) **Alternating axis of symmetry :** A molecule possesses an n-fold alternating axis of symmetry if, when rotated through an angle of 360°/n about this axis and then followed by reflection in a plane perpendicular to the axis, the molecule is indistinguishable from the original molecule.

This contains a four-fold alternating axis of symmetry. Rotation of (a) through 90° about axis AB which passes through the centre of the ring perpendicular to its plane gives (b), and reflection of (b) in the plane of the ring gives (a).

It also happens that this molecule possesses two vertical planes of symmetry (through each diagonal of the ring), but if the methyl groups are replaced alternately by the chiral groups (+)—$CH(CH_3)C_2H_5$ and (–)—$CH(CH_3)C_2H_5$, represented by Z^+ and Z^- respectively, the resulting molecule now has no planes of symmetry. Nevertheless, this molecule is not optically active since it does possess a four-fold alternating axis of symmetry [reflection of (d) (which is produced by rotation of (c) through

90° about the vertical axis) in the plane of the ring gives (c); it should be remembered that the reflection of a (+)-form is the (–)-form].

The cyclobutane derivative (c) given above to illustrate the meaning of an alternating axis of symmetry is an imaginary molecule. No compound was known in which the optical inactivity was due to the existence of only an alternating axis until McCasland and Proskow (1956) prepared such a molecule for the first time. This is a spiro-type of molecule, viz., 3, 4, 3′, 4′-tetramethylspiro-(1, 1′) dipyrrolidinium p-toluenesulphonate, (I) (the p-toluenesulphonate ion has been omitted). This molecule is discussed in some detail but here we shall examine it for its alternating axis of symmetry. Molecule (I) is superimposable on its mirror image and hence is not optically active. It does not contain a plane or centre of symmetry, but it does contain a four-fold alternating axis of symmetry. To show the presence of this axis, if (I) rotated through 900 about the co-axis of both rings, (II) is obtained. Reflection of (II) through the central plane (*i.e.*, through the N atom) perpendicular to this axis gives a molecule identical and coincident with (I).

McCasland *et al.*, (1959) have now prepared a, second compound, a pentaerythritol ester, whose optical inactivity can be attributed only to the presence of a four fold alternating axis of symmetry:

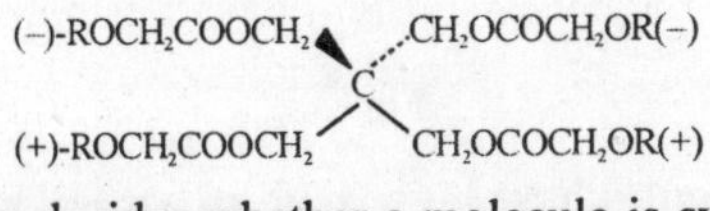

In practice one decides whether a molecule is symmetric or not by looking only for a plane or centre of symmetry, since no natural compound has yet been found to have an alternating axis of symmetry. The presence of two or more asymmetric carbon atoms will definitely give rise to optical isomerism, but nevertheless some isomers may not be optically active because these molecules as a whole are not asymmetric.

Molecular symmetry: In the above account, the test of determining whether a molecule is optically active has been to show the absence of the three elements of symmetry, (i), (ii), and (iii). We shall now consider this problem in some more detail. Symmetry problems are solved by mathematical methods known as Group Theory. A complete set of symmetry elements in any given molecule is known as a point group, *i.e.*, a point group describes the type of symmetry to which the molecule belongs. A point group is one example of groups which form the basis of group theory.

The symmetry of a molecule can be completely described in terms of symmetry elements, and the operations carried out to ascertain the presence of symmetry elements are known as symmetry operations. The basic operations are rotations and reflections. A symmetry operation may be defined as an operation which results in the conversion of the molecule into an equivalent configuration, *i.e.,* the molecule obtained after the operation is indistinguishable from the original molecule. As far as molecules are concerned, a symmetry element is a point, line or plane with respect to which one or more symmetry operations are performed. There are four basic kinds of symmetry elements, and each of these is designated by a symbol which is also used to represent the corresponding symmetry operation.

Axis of symmetry (C_n) : A C_n axis of symmetry is an axis about which the molecule can be rotated by 360°/n (2π/n rad) and thereby produce a molecule indistinguishable from the original molecule (rotation is usually taken as clockwise). The subscript n indicates the order of the axis, *i.e.,* the largest value of n for which the rotation through 360°/n produces an equivalent configuration. Some values of n (for the vertical axis) are as shown.

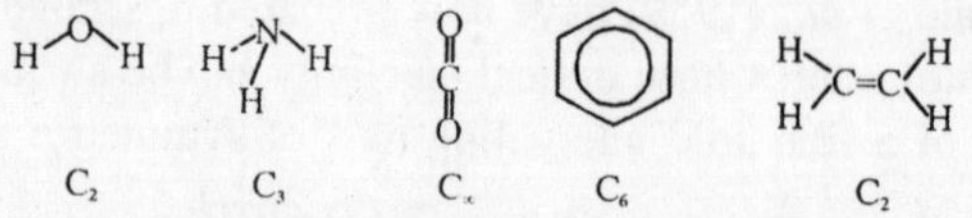

Fig. 3

All linear molecules have a C_∞ axis; an equivalent configuration is always obtained whatever is the angle of rotation. Benzene possesses a C_6 axis perpendicular to the plane of the ring; it also has six C_2 axes in the plane of the ring. Ethylene has three C_2 axes, one collinear with the C–C double bond, the second perpendicular to the plane of the molecule and passing through the centre of the double bond, and the third perpendicular to the first two and passing through the centre of the double bond.

When the value of n is one, the axis is a C_1 axis. The C_1 symmetry operation is carried out by rotating the molecule through 360°. The result is a molecule identical with the original molecule. The C_1 axis is said to be a trivial axis; all molecules possess a trivial axis. Also, every molecule possesses an identity of symmetry (E), which is observed by an identity operation (E). The operation can be carried out in various ways, *e.g,* by rotation through 360°, *i.e.,* one identity element of symmetry is C_1, a trivial axis. Since, *e.g,* C_3 represents the operation of rotation by 120° (2 × 2π/3 rad) about the C_3 axis, repetition of this operation effects the overall rotation of 240° (2 × 2π/3 rad), and another repetition effects the overall rotation of 360° (2π rad). Each result may be indicated by the symbol C_3, C_3^2, and C_3^3, respectively. At the end of the last operation, the molecule is identical with the original molecule. Hence $C_3^3 = E$.

The C_n axis is known as a proper axis; only one or more rotations about the axis are involved. When the molecule contains two or more axes of the same order, they are usually differentiated by superscript dashes, *e.g,* C_2, C'_2, and C''_2, etc. If two or more are equivalent, this is indicated by use of the same superscript dash, *e.g,* C_2, two C'_2, and two C''_2.

It is important to note that symmetry operations involving rotations are applied to the whole molecule. Since rotations of one part of the molecule with respect to another bring about changes in conformation strictly speaking the application of symmetry operations is to molecular conformations and not to 'molecules'. This may be illustrated with ethylene dichloride (see Fig. 2.11). The staggered form has a centre of symmetry, but not the fully eclipsed form. Hence, the study of molecular symmetry is the study of the molecule in a particular conformation.

Plane of symmetry (σ) : This has been defined above [see (i)]. If we suppose that the plane of symmetry is in the xy plane (of the Cartesian co-ordinates x, y, z) then, after changing the sign of the z co-ordinate for each atom from z to –z, the configuration of the molecule is equivalent to that of the original molecule. It also follows that repetition of the operation a results in the original molecule. Hence $\sigma^2 = E$. It might also be noted that the reflection plane contains a C_2 axis ($C_2^2 = E$).

When the C_n axis with the largest order (n is the greatest) is regarded as being vertical (coincident with the z-axis), planes of symmetry which are also vertical are indicated by the subscript v, *i.e.,* σ_v, If the reflection plane is in the plane of the paper, this plane of symmetry is indicated by σ_v, and if it is perpendicular to the plane of the paper, then by σ'_v. When the reflection plane is horizontal (*i.e.,* in the xy plane; C_n axis coincident with the z-axis), the plane is represented as σ_h. When a reflection plane is diagonal, *i.e.,* bisects the angles between two equivalent axes, this is indicated by σ_d.

Centre of symmetry (i) : This has been defined above [see (ii)]. If we use the Cartesian co-ordinate system, it can be seen that if the molecule has a centre of symmetry, then changing the co-ordinates (x, y, z) of every atom, with this centre as origin, to (–x, –y, –z) produces an equivalent configuration of the molecule. This operation, also denoted by i, is known as inversion and hence a centre of symmetry is also called a centre of inversion.

Alternating axis -of symmetry (S_n) : This has been defined above [see (iii)]. Since the operation S_n involves rotation followed by reflection, an alternating axis of symmetry is also called a rotationreflection axis of symmetry. This type of axis is called an improper axis; two steps are involved: rotation first (about an axis) followed by reflection (the order of operations may be reversed without affecting the result).

From what has been said above, it can be shown that $S_1 \equiv \sigma$ and $S_2 \equiv i$. The cyclobutane and spiro-N-compounds described have an S_4 axis. In this case, there is neither a plane nor a centre of symmetry present.

As we have seen, if a molecule is not superimposable on its mirror image, that molecule can exhibit optical activity. Such molecules are (a) asymmetric; these are completely devoid of any symmetry elements (except a trivial axis); or (b) dissymmetric; these have proper axes but no improper axis. It therefore follows that if a molecule has an S_n axis, that molecule is not optically active, whereas if it has no S_n axis, that molecule is optically active. Alternatively, if a C_n axis is the only symmetry element present in a molecule, that molecule is optically active.

Of all the possible point groups (see above), those of C_n and D_n contain only proper axes of rotation as their symmetry element. Hence, only molecules belonging to these groups are capable of exhibiting optical activity.

For our purpose, we may define a C_n point group as one which contains the symmetry element C_n. A D_n point group is one which contains the symmetry elements C_n and n C_2 axes. The C_2 axes are all perpendicular to the C_n axis and make equal angles with each other. The D_n point group is also known as dihedral symmetry (n is the principal axis). All other point groups contain at least one of the symmetry elements S_n, σ, or i, *e.g,* an S_n point group contains the symmetry element S_n, a C_{nv} point group contains a C_n axis of symmetry and nσ planes of symmetry, all of which contain C_n.

Optical Isomerism

In optical isomerism we have a much more subtle phenomenon than even the geometrical isomcrisra. While the geometrical isomers differ in physical properties such as melting point, boiling point, density etc., and also in certain chemical properties, the optical isomers will have the same chemical reactions and will be alike in all physical properties mentioned above. They can only be distinguished by their *'action on plane-polarized light'*. This property which is often referred to as the *Optical activity* requires a brief discussion.

Thus, in ordinary light vibrations occur in all planes at right angles to the line of propagation. In plane polarized light the vibrations tak place only in one plane, vibrations in other planes being cut off. Plane polarized light can be obtained by passing ordinary light through a Nicol prism.

Certain organic compounds, when their solutions are placed in the path of a plane polarized light, have the remarkable property of rotating its plane through a certain angle which may be either to the left or to the right. If the polarized light has its vibrations in the plane XY before entering such a solution, the direction on leaving it will be changed to say X'Y', the plane having been rotated through the angle A given molecule which has at least one element of symmetry when its 'classical' configuration (*i.e.,* the Fischer projection formula) is inspected may, however, have a conformationwhich is devoid of any element of symmetry. At first sight, such a molecule might he supposed to be optically active.

In practice, however, it is not; individual molecules are optically active, but statistically, the whole collection of molecules is not. It therefore follows that when a molecule can exist in one or more conformations, then provided that at least one of the conformations (whether preferred or not) is superimposable on its mirror image, the compound will not be optically active.

OPTICAL ACTIVITY DUE TO CRYSTALLINE STRUCTURE

here are many substances which are optically active in the solid state only, *e.g,* quartz, sodium chlorate, benzil, etc. Let us consider quartz, the first substance shown to be optically active (Arago, 1811). Quartz exists in two crystalline forms, one of which is dextrorotatory and the other laevorotatory. These two forms are mirror images and are not superimposable. Such pairs of crystals are said to be enantiomorphous (quartz crystals are actually hernihedral and are mirror images). X-ray analysis has shown that the quartz crystal lattice is built up of silicon and oxygen atoms arranged in left- and right-handed spirals. One is the mirror image of the other, and the two are not superimposable. When quartz crystals are fused, the optical activity is lost. Therefore the optical activity is entirely due to the asymmetry of the crystalline structure, since fusion brings about only a physical change. Thus we have a group of substances which are optically active only so long as they remain solid; fusion, vaporisation or solution in a solvent causes loss of optical activity.

Optical activity due to molecular structure : There are many compounds which are optically active in the solid, fused, gaseous or dissolved state, *e.g,* glucose, tartaric acid, etc. In this case the optical activity is entirely due to the asymmetry of the molecular structure. The original molecule and its non-superimposable mirror image are known as enantiomorphs (this name is taken from crystallography), enantiomers, or optical antipodes.

Properties of enantiomers : It appears that enantiomers are identical physically except in two respects:

(i) their manner of rotating polarised light; the rotations are equal but opposite.

(ii) the absorption coefficients for dextro- and laevocircularly polarised light are different; this difference is known as circular dichroism.

The crystal forms of enantiomers may be mirror images of each other, *i.e.,* the crystals themselves may be enantiomorphous, but this is unusual. Enantiomers are similar chemically, but their rates of reaction with other optically active compounds are usually different. They may also be different physiologically, *e.g,* (+)-histidine is sweet, (–)-tasteless; (–)-nicotine is more poisonous than (+). The mass spectra of enantiomers (and their corresponding racemates) are identical, and so are their NMR spectra.

THE TETRAHEDRAL CARBON ATOM

In 1874, van't Hoff and Le Bel, independently, gave the solution to the problem of optical isomerism in organic compounds. van't Hoff proposed the theory that if the four valencies of the carbon atom are arranged tetrahedrally (not necessarily regular) with the carbon atom at the centre, then all the cases of isomerism known are accounted for.

Le Bel's theory was substantially the same as van't Hoff's, but differed in that whereas van't Hoff believed that the valency distribution was definitely tetrahedral and fixed as such, Le Bel believed that the valency directions were not rigidly fixed, and did not specify the tetrahedral arrangement, but thought that whatever the spatial arrangement, the molecule Cabde would be asymmetric. Later work has shown that van't Hoff's theory is more in keeping with the facts.

Both van't Hoff's and Le Bel's theories were based on the assumption that the four hydrogen atoms in methane are equivalent; this assumption has been shown to be correct by means of chemical and physico-chemical methods. Before the tetrahedral arrangement was proposed, it was believed that the four carbon valencies were planar, with the carbon atom at the centre of a square (Kekule, 1858).

Pasteur (1848) stated that all substances fell into two groups, those which were superimposable on their mirror images, and those which were not. In substances such as quartz, optical activity is due to the dissymmetry of the crystal structure, but in compounds like sucrose the optical activity is due to molecular dissymmetry. Since it is impossible to have molecular dissymmetry if the molecule is flat, Pasteur's work is based on the idea that molecules are three-dimensional and arranged dissymmetrically.

A further interesting point in this connection is that Pasteur quoted an irregular tetrahedron as one example of a dissymmetric structure. Also, Paterno (1869) had proposed tetrahedral models for the structure of the isomeric compounds $C_2H_4Cl_2$ (at that time it was thought that there were three isomers with this formula; one ethylidene dichloride and two ethylene dichlorides).

Evidence for the Tetrahedral Carbon Atom

The molecule CX_4 constitutes a five-point system, and since the four valencies of carbon are equivalent, their disposition in space may be assumed to be symmetrical. Thus there are three symmetrical arrangements possible for the molecule CX_4, one planar and two solid-pyramidal and tetrahedral. By comparing the number of isomers that have been prepared for a given compound with the number predicted by the above three spatial arrangements, it is possible to decide which one is correct.

Compounds of the types Ca_2b_2 and Ca_2bd. Both of these are similar, and so we shall only discuss molecule Ca_2b_2.

(i) If the molecule is planar, then two forms are possible. This planar configuration can be either square or rectangular; in each case there are two forms only.

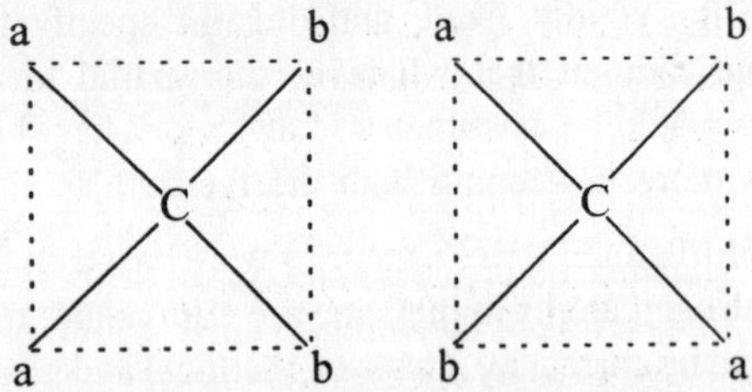

(ii) If the molecule is pyramidal, then two forms are possible. There are only two forms, whether the base is square or rectangular.

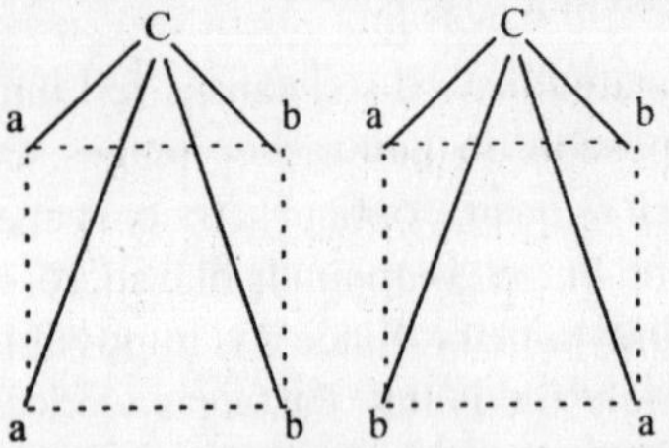

(iii) If the molecule is tetrahedral, then only one form is possible the carbon atom is at the centre of the tetrahedron).

In practice, only one form is known for each of the compounds of the types Ca_2b_2 and Ca_2bd; this agrees with the tetrahedral configuration.

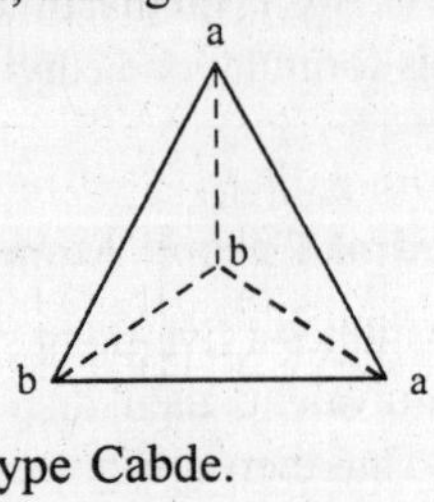

Compounds of the type Cabde.

(i) If the molecule is planar, then three forms are possible.

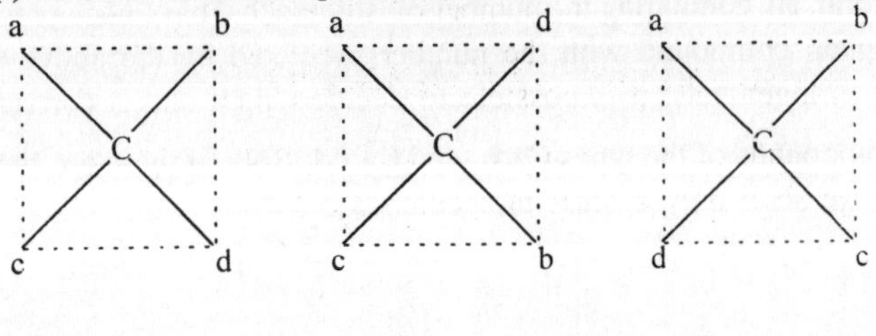

(ii) If the molecule is pyramidal, then six forms are possible; there are three pairs of enantiomers. Each of the forms drawn as a pyramid, is not superimposable on its mirror image, *e.g,* pair of enantiomers.

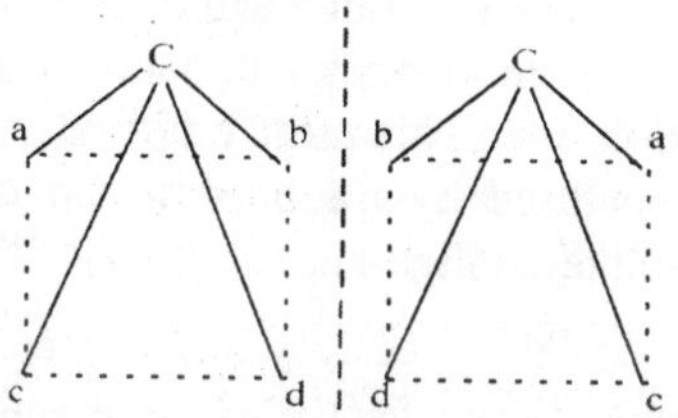

(iii) If the molecule is tetrahedral, there are two forms possible, one related to the other as object and mirror image, which are not superimposable, *i.e.,* the tetrahedral configuration gives rise to one pair of enantiomers.

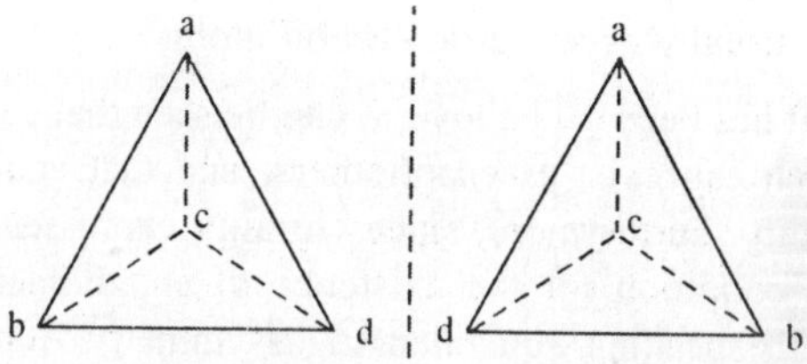

In practice, compounds of the type Cabde give rise to only one pair of enantiomers; this agrees with the tetrahedral configuration.

When a compound contains four different groups attached to a carbon atom, that carbon atom is said to be asymmetric (actually, of course, it is the group which is asymmetric; a carbon atom cannot be asymmetric). The majority of optically active compounds (organic) contain one or more asymmetric carbon atoms. It should be remembered, however, that the essential requirement for optical activity is the asymmetry of the molecule. A molecule may contain two or more asymmetric carbon atoms and still not be optically active.

A most interesting case of an optically active compound containing one asymmetric carbon atom is the resolution of s-butylmercuric bromide, EtMeCHHgBr (Hughes, Ingold *et al.,* 1958). This appears to be the first example of the resolution of a simple organometallic compound where the asymmetry depends only on the carbon atom attached to the metal.

Chirality

As we have seen, the structures of enantiomers differ only in 'handedness', one being left-handed and the other being right-handed. Any molecule which is not superimposable on its mirror image is said to possess chirality (from the Greek kheir, hand). Thus, the term chirality means 'having handedness'. This term was first introduced by Kelvin (1884), and has been used by Cahn *et al.,* in their system for the specification of absolute configuration.

Chirality expresses the necessary and sufficient condition for the existence of enantiomers. The adjective chiral is equivalent to being left- or right-handed, and so a chiral centre is one which can be left- or right-handed. On the other hand, when a molecule is superimposable on its mirror image, that molecule does not possess 'handedness' (*i.e.,* it is not asymmetric) and is said to be achiral. The commonest cause of optical activity is the presence of one or more chiral centres which, in organic chemistry, are usually asymmetric carbon atoms.

From what has been said above, it can be seen that chiral molecules are those which can exist as enantiomers, and that enantiomers have opposite chirality. Furthermore, since chirality expresses the necessary and sufficient condition for the existence of enantiomers, chirality is therefore, strictly speaking, equivalent to dissymmetry. Achiral molecules are symmetric or non-dissymmetric (the latter term is preferred by many authors).

Isotopic Asymmetry

In the optically active compound Cabde, the groups a, b, d and e (which may or may not contain carbon) are all different, but two or more may be structural isomers, *e.g,* propylisopropylmethanol is optically active. The substitution of hydrogen by deuterium has also been investigated in recent years to ascertain whether these two atoms are sufficiently different to give rise to optical isomerism. The earlier work gave conflicting results, but later work, however, is definitely conclusive in favour of optical activity, *e.g,* Eliel (1949) prepared optically active phenylmethyldeuteromethane, $CH_3CHDC_6H_5$, by reducing optically active phenylmethylmethyl chloride (x-phenylethyl chloride), $CH_3CHClC_6H_5$, with lithium aluminium deuteride; Ross *et al.,* (1956) have prepared (–)-2-deuterobutane by reduction of (–)-2-chlorobutane with lithium aluminium deuteride; and Alexander *et al.,* (1949) reduced trans-2-pmenthene with deuterium (Raney nickel catalyst) and obtained a 2,3-dideutero-trans-p-menthane (I) that was slightly laevorotatory. Alexander (1950) also reduced (–)-menthyl toluene-p-sulphonate and obtained an optically active 3deutero-trans-p-menthane (II).

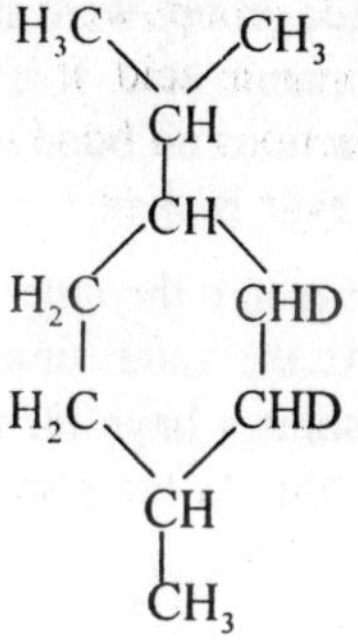

(I)

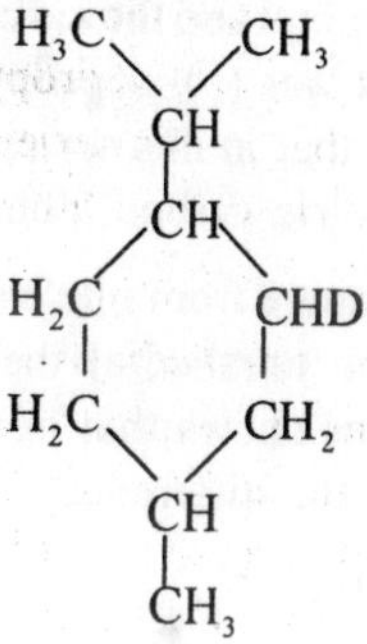

(II)

Some other optically active compounds with deuterium asymmetry are, *e.g,* (III; Streitwieser, 1955) and (IV; Levy *et al.*, 1957):

$CH_3CH_2CH_2CHDOH$ (III) $\quad$ CH_3CHDOH (IV)

A point of interest here is that almost all optically active deuterium compounds have been prepared from optically active precursors. Exceptions are (V) and (VI), which have been resolved by Pocker (1961).

$C_6H_5CHOHC_6D_5$ (V) $\quad$ $C_6H_5CDOHC_6D_5$ (VI)

It might be noted that chirality produced by replacement of hydrogen by deuterium is of two types:

(i) deuterium is directly attached to the chiral centre (asymmetric carbon atom), *e.g,* (III) and (IV);

(ii) deuterium is not directly attached to the chiral centre, *e.g,* (V).

Further evidence for the tetrahedral carbon atom

(i) Conversion of the two forms (enantiomers) of the molecule Cabde into Ca_2bd results in the formation of one compound only (and disappearance of optical ––tivity), *e.g,* both dextro- and laevorotatory lactic acid may be reduced to the sane propionic acid, which is not optically active. These results are possible only with a tetrahedral arrangement.

(ii) If the configuration is tetrahedral, then interchanging any two groups in the molecule Cabde will produce the enantiomer, *e.g,* b and e. Fischer and Brauns (1914), starting with (+)-isopropylmalonamic acid, carried out a series of reactions whereby the carboxyl and the carbonamide groups were interchanged: the product was (–)-isopropylmalonamic acid. It is most important to note that in this series of reactions no bond connected to the asymmetric carbon atom was ever broken.

This change from one enantiomer into the other is in agreement with the tetrahedral theory. At the same time, this series of reactions shows that optical isomers have identical structures, and so the difference must be due to the spatial arrangement.

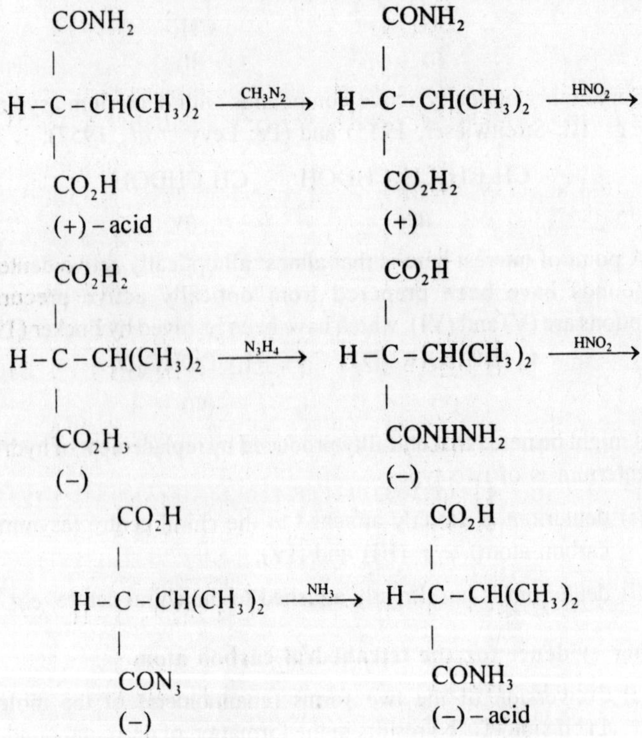

(iii) X-ray crystallography, dipole moment measurements, absorption spectra and electron diffraction studies show that the four valencies of carbon are arranged tetrahedrally with the carbon atom inside the tetrahedron.

It should be noted in passing that the tetrahedra are not regular unless four identical groups are attached to the central carbon atom; only in this case are the four bond lengths equal. In all other cases the bond lengths will be different, the actual values depending on the nature of the atoms joined to the carbon atom.

TWO POSTULATES UNDERLIE THE TETRAHEDRAL THEORY

(i) **The principle of constancy of the valency angle :** Mathematical calculation of the angle subtended by each side of a regular tetrahedron at the central carbon atom gives a value of 109° 28′. Originally, it was postulated (van't Hoff) that the valency angle was fixed at this value. It is now known, however, that the valency angle may deviate from this value. The four valencies of carbon are formed by hybridisation of the $2s^2$ and $2p^2$ orbitals, *i.e.,* there are four spa bonds. Quantum mechanical calculations show that the four carbon valencies in the molecule Ca_4 are equivalent and directed towards the four corners of a regular tetrahedron.

Furthermore, quantum-mechanical calcula-tions require the carbon bond angles to be close to the tetrahedral value, since change from this value is associated with loss in bond strength and consequently decrease in stability. According to Coulson *et al.,* (1949), calculation has shown that the smallest valency angle that one can reasonably expect to find is 104°. It is this value which is found in the cyclopropane and cyclobutane rings, these molecules being relatively unstable because of the 'bent' bonds.

(ii) **The principle of free rotation about a single bond:** Originally, it was believed that internal rotation about a single bond was completely free. Let us consider the ethane molecule, CH_3—CH_3, and let us imagine that one methyl group is rotated about the C–C bond as axis with the other group at rest. Suppose we use, as the starting point, the position in which two C-H bonds are parallel, *i.e.,* these four atoms lie in a plane.

In this position, the dihedral angle (angle of rotation or angle of torsion) is zero, and if the rotation is free, then as the dihedral angle changes, the energy content of the molecule will remain constant (the plot of energy content against the dihedral angle will be a horizontal line). In this situation, the two 'halves' can assume, with complete freedom, an infinite number of positions relative to each other. Thus, the entropy of the molecule will be a maximum. Pitzer *et al.*, (1936) calculated the entropy of ethane based on the assumption that the internal rotation was free, and found that calculated value was greater than the observed value. This means that there is less freedom to assume all possible dihedral angles than was expected on the principle of free rotation. Pitzer therefore assumed that the internal rotation is hindered by a potential energy barrier, and calculated the change in entropy with increasing barrier height.

When he assumed a barrier of 12.55 kJ mol^{-1} (3 kcal), the calculated and observed entropies were brought into good agreement. The potential energy curve obtained for ethane is shown in Fig. 2.10a and, by convention, the potential energy is measured relative to the energy of the most stable form. Because of the existence of potential energy barriers, the result is mutual oscillation (libration) about the conformations with minimum potential energy, thereby producing 'more order' (more restriction) than had there been no barriers.

The Newman projection formula. This is obtained by viewing the molecule along the bonding line of the two carbon atoms, with the carbon atom nearer to the eye being designated by equally spaced radii, and the carbon atom further from the eye by a circle with three equally spaced radial extensions. It can be seen from Fig. 2.10a that the energy of the eclipsed conformation is greater than that of the staggered conformation. The potential energy barrier (11.92 kJ mol^{-1}) is much too small for either conformation to remain stable, *i.e.*, the eclipsed and staggered forms are readily interconvertible and hence neither can be isolated as such. However, the staggered conformation is the preferred form, *i.e.*, its population is greater than that of the eclipsed form.

Now let us consider the case of ethylene dichloride. According to Bernstein (1949), the potential energy of ethylene dichloride undergoes the changes shown in Fig. 2.11 when one CH_2Cl group is rotated about the C–C bond with the other CH_2Cl at rest. There are two positions of minimum energy, one corresponding to the staggered (transoid or anti) form and the other to the gauche (skew) form, the latter possessing approximately 4.6 kJ more than the former. The fully eclipsed (cisoid) form possesses about 18.83 kJ more energy than the staggered form and thus the latter is the preferred form, *i.e.,* the molecule is largely in this form.

Dipole moment studies show that this is so in practice, and also show (as do Raman spectra studies) that the ratio of the two forms varies with the temperature. Furthermore, infrared, Raman spectra and electron diffraction studies have shown that the gauche form is also present. According to Mizushima *et al.*, (1938), only the staggered form is present at low temperatures. The problem of internal rotation about the central C–C bond in n-butane is interesting, since the values of the potential energies of the various forms have been used in the study of cyclic compounds. The various forms are if the energy content of the staggered form is taken as zero, then the other forms have the energy contents shown (Pitzer, 1951).

From the foregoing account it can be seen that in theory, there is no free rotation about a single bond. In practice, however, it may occur if the potential barriers of the various forms do not differ by more than about 40 kJ mol^{-1}. Free rotation about a single bond is generally accepted in simple molecules. Restricted rotation, however, may occur when the molecule contains groups large enough to impede free rotation, *e.g,* in ortho-substituted biphenyls. In some cases resonance can give rise to restricted rotation about a 'single' bond.

In addition to the nomenclature of conformers described above-staggered (anti), skew (gauche), etc., conformers are also described in terms of the dihedral angle between specified groups. Thus, the dihedral angle for the skew form of n-butane (2 and 2′,) is 60°; for the staggered (anti) form the dihedral angle is 180°.

Table 1

Dihedral angle	*Designation*	*Symbol*
– 30° to + 30°	± syn-periplanar	± SP
+ 30° to + 90°	+ syn-clinal	+sc
+ 90° to + 150°	+ anti-clinal	+ac
+ 150° to – 150°	± anti-periplanar	± ap
– 30° to – 90°	– syn-clinal	– sc
– 90° to – 150°	– anti-clinal	– ac

In many cases, however, the exact dihedral angle is not known, and to describe these cases, the following nomenclature in terms of approximate dihedral angles has been proposed (Klyne and Prelog, 1960). The terms syn and anti are used to indicate respectively a dihedral angle smaller or greater than 90°. The terms 'periplanar' and 'clinal' respectively describe approximately planar (0° ± 30° and 180° ± 30°) and inclined positions (all other angles).

This method of nomenclature is summarised in the diagram.

CONFORMATIONAL ANALYSIS

Isomers which are formed by rotation about single bonds are called different conformations or conformers. The terms rotational isomers and constellations have also been used in the same sense as conformations.

Various definitions have been given to the term conformation (which was originally introduced by W. N. Haworth, 1929). In its widest sense, conformation has been used to describe different spatial arrangements of a molecule which are not superimposable. This means, in effect, that the terms conformation and configuration are equivalent. The definition of configuration, in the classical sense, does not include the problem of the internal forces acting on the molecule.

The term conformation, however, is the spatial arrangement of the molecule when all the internal forces acting on the molecule are taken into account. In this more restricted sense, the term conformation is used to designate different spatial arrangements arising by twisting or rotation of bonds of a given configuration (used in the classical sense).

The existence of potential energy barriers between the various conformations shows that there are internal forces acting on the molecule. The nature of these interactions that prevent free rotation about single bonds, however, is not completely clear. According to one theory, the hindering of internal rotation is due to dipole-dipole forces. Calculation of the dipole moment of ethylene dichloride on the assumption of free rotation gave a value not in agreement with the experimental value. Thus free rotation cannot be assumed, but on the assumption that there is interaction between the two groups through dipole-dipole attractive or repulsive forces, there will be preferred conformations, *i.e.,* the internal rotation is not completely free.

This restricted rotation is shown by the fact that the dipole moment of ethylene dichloride increases with temperature; in the staggered form the dipole moment is zero, but as energy is absorbed by the molecule, rotation occurs to produce finally the eclipsed form in which the dipole moment is a maximum. Further work, however, has shown that factors other than dipole-dipole interactions must also be operating in opposing the rotation. One of these factors is steric repulsion, *i.e.*, repulsion between the non-bonded atoms (of the rotating groups) when they are brought into close proximity.

The existence of steric repulsion may be illustrated by the fact that although the bond moment of C–Cl is greater than that of C–Br, the energy difference between the eclipsed and staggered conformations of ethylene dichloride is less than that of ethylene dibromide. Furthermore, if steric repulsion does affect internal rotation, then in the ethylene dihalides, steric repulsion between the hydrogen and halogen atoms, if sufficiently large, will give rise to two other potential energy minima.

Other factors also affect stability of the various conformations. Staggered and skew forms always exist in molecules of the type CH_2Y–CH_2Z (where Y and Z are Cl, Br, I, CH_3, etc.), and usually the staggered form is more stable than the skew. In a molecule such as ethylene chlorohydrin or ethylene glycol, however, intramolecular hydrogen bonding is possible in the skew form but not the staggered. This would stabilise the molecule by about 20-29 kJ mol^{-1}, and this is great enough to make the skew form more stable than the staggered. Infrared spectroscopy has shown that the skew form predominates. In addition to the factors already mentioned, there appear to be other factors that cause the absence of complete free rotation about a single bond, *e.g*, the energy barrier in ethane is too great to be accounted for by steric repulsion only.

Several explanations have been offered; *e.g*, Pauling (1958) has proposed that the energy barrier in ethane (and in similar molecules) results from repulsions between adjacent bonding pairs of electrons, *i.e.*, the bonding pairs of the C–H bonds on one carbon atom repel those on the other carbon atom. Thus the preferred conformation will be the staggered. It is still possible, however, that steric repulsion is also present, and this raises the barrier height.

When the stability of a molecule is decreased by internal forces produced by interaction between constituent parts, that molecule is said to be under strain. In view of the foregoing discussion, it can be seen that there are four contributing sources to strain:

(i) steric strain,

(ii) dipole-dipole interactions,

(iii) bond angle strain, and

(iv) bond opposition strain.

Which of these plays the predominant part depends on the nature of the molecule in question. This study of the existence of preferred conformations in molecules, and the relating of physical and chemical properties of a molecule to its preferred conformation, is known as conformational analysis. The energy differences between the various conformations determine which one is the most stable, and the ease of transformation depends on the potential energy barriers that exist between these conformations.

It should be noted that the molecule, in its unexcited state, will exist largely in the conformation of lowest energy content. If, however, the energy differences between the various conformations are small, then when excited, the molecule can take up a less favoured conformation, *e.g,* during the course of reaction with other molecules.

So far, we have considered the conformations of saturated compounds. Conformational studies of unsaturated compounds and compounds containing the oxo group have led to some unexpected results, *e.g,* microwave spectroscopy has shown that the preferred conformations of propene (Herschbach *et al.,* 1958) and acetaldehyde (Kilb *et al.,* 1957) are the eclipsed forms, and NMR spectroscopy has shown that the predominant conformation of propionaldehyde is the one in which the methyl group and oxygen atom are eclipsed (Pople *et al.,* 1960). The reason for these observations is uncertain. Many methods are now used to investigate the conformations of molecules: thermodynamic calculations, dipole moments, X-ray and electron diffraction, infrared and ultraviolet spectroscopy, chemical methods, etc.

A particularly useful method for studying the conformations of molecules is NMR spectroscopy. However, before dealing with this, let us first consider the chemical shifts of the protons of CH_3, CH_2, and the CH group in acyclic compounds and the chemical shift of a proton attached to O, N, etc. As we have seen the more electronegative Z is in the groups –CH–Z and –Z–H, the more is the proton deshielded and consequently the lower is the r-value.

Let us now consider ethyl chloride, and since the most stable conformation is the staggered one, the molecule may be represented as (I), (II), and (III). In (I), the environments of H_a and H_c are identical, but differ from that of H_b. This can be seen to be the case by replacing each proton, one at a time, by Z. This procedure for H_a and H_c produces mirror images, and so the two protons are chemically equivalent, but for Hb the substituted product is different, and therefore H_b is not chemically equivalent to H_a and H_c. The coupling constants of H_a with H_d and H_e are different because of the different dihedral angles, and similarly for H_c.

Thus, H_a and H_c are chemically but not magnetically equivalent. By using similar arguments, it can be shown that H_d and H_c are also chemically but not magnetically equivalent. Hence, if the population of ethyl chloride conformation were completely represented by (I), protons Ha and. He would give one signal and Hb another signal provided that the chemical shifts were sufficiently different (0.1–0.2 p.p.m.), *i.e.*, the methyl group would give two signals. In practice, the methyl group gives only one signal (a triplet), and therefore protons H_a, H_b, and H_c must be equivalent. This is explained on the basis that there is free rotation about the carbon-carbon single bond. When the methyl group rotates (with respect to the CH_2Cl group), it can take up the other two stable staggered conformations, (II) and (III), the result being that each proton has an average environment. Since these average environments are identical, rotation results in the equivalence of H_a, H_b, and H_c. This equivalence may be demonstrated as follows. Let P_I, P_{II}, P_{III} be the populations in conformations (I), (II), and (III), respectively, where P_I, P_{II}, P_{III} = 1.

Also, let δ (the chemical shift) of any proton trans (anti) to another proton be x, and y if trans to Cl. Now, the observed δ of a given proton is the sum of the δ-contributions in each conformation, and so it follows that:

$$\delta(H_a) = P_I x + P_{II} y + P_{III} x$$

$$\delta(H_b) = P_I y + P_{II} x + P_{III} x$$

$$\delta(H_c) = P_I x + P_{II} x + P_{III} y$$

Since all three populations are equal (all three are indistinguishable because all protons, as such, are identical), *i.e.*, $P_I = P_{II} = P_{III} = 1/3$, therefore

$$\delta(H_a) = \delta(H_b) = \delta(H_c) = 2/3x + 1/3y$$

Hence, all three protons have the same (average) chemical shift (in the freely rotating state) and so there is no observable splitting for the methyl group.

Application of this method to protons H_d and H_e gives:

$$\delta(H_d) = P_I x + P_{II} x + P_{III} x = x$$

$$\delta(H_e) = P_I x + P_{II} x + P_{III} x = x$$

From the above discussion, it can be seen that if ethyl chloride were undergoing slow rotation, it would be an ABB′CC′ spin system, but when undergoing fast rotation, it is an A_3B_2 spin system. Now let us consider the case of 1-bromo-2-chloroethane, and if we use the same arguments as before, then in (IV), H_a and H_b are chemically but not magnetically equivalent, and this is also true for H_c and H_d. Also, if δ for a proton trans to a proton is x, trans to Cl is y, and trans to Br is z, then

$$\delta(H_a) = P_{IV} x + P_V y + P_{VI} x$$

$$\delta(H_b) = P_{IV} x + P_V x + P_{VI} Y$$

In this molecule, although $P_V = P_{VI} \neq P_{IV}$, it still follows that $\delta(H_a) = \delta(H_b)$. In the same way, it can be shown that $\delta(H_c) = \delta(H_d)$.

We can extend the argument by also considering the possible coupling constants. If J, and Js represent, respectively, the vicinal trans and skew coupling, then:

$$J_{ad} = P_{IV} J_t + P_V J_s + P_{VI} J_s$$

$$J_{bd} = P_{IV} J_s + P_V J_t + P_{VI} J_s$$

Since $P_V = P_{VI} \neq P_{IV}$, therefore $J_{ad} \neq J_{bd}$. Hence, H_a and H_b (and also H_c and H_d) are chemically but not magnetically equivalent. Application of this method will show that the three protons in the methyl group in ethyl chloride are magnetically equivalent, as are also the two protons in the CH_2Cl group, for the freely rotating molecule.

Finally, let us consider a molecule of the type $R^1R^2CHCHR^3R^4$. Each carbon atom is asymmetric, and the environments of H_a are different in all three conformations, and this is also true for H_b.

Furthermore, because of the different steric effects, the three populations will be different (*i.e.*, $P_{VII} \neq P_{VIII} \neq P_{IX}$). Hence, the chemical shifts of H_a (and those of H_b) are different in each conformation. If rotation were slow, the NMR spectrum would be a composite of all three spectra (three AB spizz systems for each pair of enantiomers). If the temperature is raised, the rate of rotation is increased, and the result is an average of chemical shifts and coupling constants to give one AB spin system (for each pair of enantiomers). These averages, however, will be weighted in favour of the most stable conformation, *i.e.,* the one with the highest population. Now let us consider the case of amides. Infrared spectroscopic evidence has indicated that amides are resonance hybrids, and this' is supported by NMR studies, *e.g,* the two methyl groups in dimethylformamide give two separate signals at room temperature. This can be explained on the basis that, because of the partial double bond character of the C–N bond, the two methyl groups have different

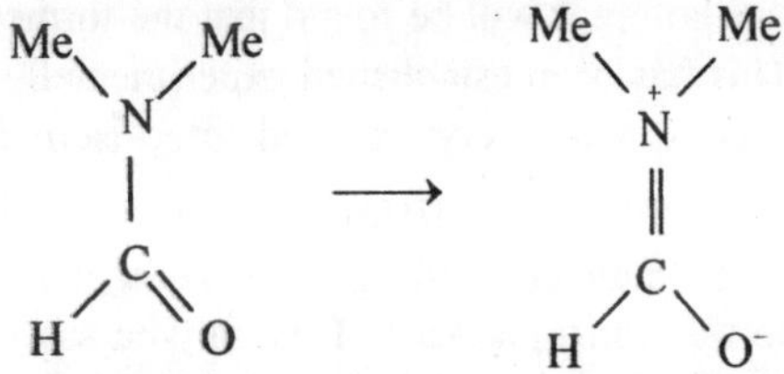

environments, one methyl group being cis and the other trans to the hydrogen atom. At room temperature the rate of rotation about the C–N bond is slow enough for the two methyl groups to be non-equivalent. As the temperature is raised, the signals broaden and finally collapse to a single line.

At these higher temperatures, the molecule has absorbed sufficient energy to overcome the energy barrier to rotation to make the average environment of the two methyl groups the same, *i.e.,* the two methyl groups are now equivalent.

DIFFERENCES IN STABILITY AND REACTIVITY OF DIASTEREOISOMERS

The stabilities of diastereoisomers of compounds containing two asymmetric carbon atomsare generally different, but these differences are usually small. The meso-form is generally more stable than either of the active forms, and the erythro compounds are generally more stable than the threo compounds. This may be demonstrated by consideration of the molecule CSML–CSML, where S, M, and L represent the smallest, medium, and largest groups, respectively. First let us consider the meso-form and one active form, each being drawn in its most stable conformation (S = S, M = M, L = L; *e.g,* $HO_2CCHOHCHOHCO_2H$).

The basis of the argument is the general principle that crossed steric interactions between groups of different size are less than the sum of the steric interactions between groups of equal size. The sum of the skew interactions in the meso-form is: 2(L: M) + 2(M: S) + 2(L: S), whereas in the active form the sum is : 2(L: M) + (M: M) + 2(L: S) + (S: S). Therefore, since (M: M + S : S) > 2(M: S), the meso-form is under less steric strain than the active form, and consequently is the more stable diastereoisomer.

If any two of the groups are not the same, *e.g,* $HO_2CCHOHCHOH$ CHO, the compound will be the erythro- or threo-isomer, and using the same arguments as before it will be found that the former is more stable than the latter. This has been established experimentally for many pairs of meso- and racemic and of erythro- and threo-isomers.

When we consider the differences in reactivity between diastereoisomers, the main controlling factor is the height of the energy barrier leading to the transition state. This may be assessed in terms of two factors : the steric factor which is concerned with the conformational requirements of groups not involved in the reaction, and the stereoelectronic factor which is concerned with the spatial relationships that exist between electrons involved in bond formation and/or bond breaking in the transition state. Acyclic systems can usually adjust themselves to the stereoelectronic requirements of the transition state, about 4.2-8.4 kJ/mol being required in the process. With cyclic systems, because of their relative rigidity, adjustment in a similar fashion may be far more difficult.

The foregoing account of the problem of conformations of molecules has been mainly qualitative. It is also of interest to consider the problem from a semi-quantitative point of view.

The most convenient parameters for defining the spatial arrangement of the atoms in a molecule are bond lengths, bond angles, and torsional (dihedral) angles, and the changes in energy content in the molecule will depend on the changes in these parameters. These changes-bond stretching (and compression), bond angle bending, and bond torsion—are collectively called molecular deformations.

Bond stretching and compression. If the potential energy of two particles in their equilibrium position, *i.e.,* when they are separated by the bond length, (r), is taken as zero, then the P.E., V_r, of the two particles when the bond length is changed by Δr is given by the expression

$$V_r = \frac{1}{2}k_r(\Delta r)^2$$

where k_r is the bond stretching force constant. For both C–C and C–H bonds, k_r is about 5 × 10^5 dynes/cm (5 N/cm) and the above expression (for these two bonds) reduces to

$$\begin{aligned} V_r &= 350(\Delta r)^2 \text{ kcal/mol/Å}^2 \\ &= 1464(\Delta r)^2 \text{ kJ/mol/Å}^2 \end{aligned}$$

With the C–C bond length equal to 1.54Å, then a change of 2 per cent, *i.e.,* 0.031Å, is equal to a change in P.E. of ~ 1.42 kJ/mol.

Bond angle bending. If the C–C–C valency angle in saturated n-hydrocarbons is taken as the standard value (~112°), then any deviation from this value produces angle strain (also known as Baeyer strain or classical strain). If the angle deformation is $\Delta\theta$, then the angle strain, V_θ, is given by the expression

$$V_\theta = \frac{1}{2}k_\theta(\Delta\theta)^2$$

where k_θ is the bond bending force constant. Since k_θ has similar values for most C–C–C bond angles, the above expression may be reduced to

$$\begin{aligned} V_\theta &= 0.01(\Delta\theta)^2 \text{ kcal/mol/deg}^2 \\ &= 0.042(\Delta\theta)^2 \text{ kJ/mol/deg}^2 \end{aligned}$$

Thus, if $\Delta\theta$ = 6°, then V_θ = 1.5 kJ/mol. This is roughly the same value as V_r for a 2 per cent change in r (see above). If we consider acyclic compounds only, then the maximum value for $\Delta\theta$ is about 10–12°. Thus, an angle deformation of 6° is about 50 per cent of the maximum value. Hence angle deformation is more easily brought about than linear deformation.

Bond torsion : As we have seen, the P.E. of the system varies with the torsional angle, $\Delta\phi$. If V_0 is the torsional energy barrier, *i.e.,* the barrier height between a maximum and a minimum, then the variation in P.E., V_ϕ, (the torsional strain or Pitzer strain) is given by the expression

$$\begin{aligned} V_\theta &= \frac{1}{2}V_0(1 + n\cos\Delta\theta) \text{ kcal/mol} \\ &= 2.1\, V_0(1 + n\cos\Delta\phi) \text{ kJ/mol} \end{aligned}$$

where n is the number of P.E. minima that occur in the rotation through 360°. This equation may be applied to molecules such as ethane (n=3).

Steric repulsion : This interaction between non-bonded atoms is also a function of the three parameters r, θ, and φ. These three may be replaced by a fourth parameter, ρ, the distance between non-bonded atoms, *i.e.,* V_ρ, the steric strain, is estimated in terms of ρ.

It can therefore be seen that molecular strain energy is the sum of the four contributing factors, V_r, V_θ, V_ϕ, and V_ρ. Furthermore, since a molecule will normally be in the state corresponding to its lowest P.E., strain energy is thus the increase in energy which is produced by deviations of the parameters from their most favourable values. Unfortunately, it is not easy to estimate the various contributions, but even so, it may be possible to obtain approximate values which may be used in judging the stabilities of various conformations. It also appears that, in general, molecules are subject to very little linear deformations. This, however, is not the case when the transition state is entered by the molecule during reaction.

CONVENTIONS USED IN STEREOCHEMISTRY

The original method of indicating enantiomers was to prefix each one by d or l according as it was dextrorotatory or laevorotatory. van't Hoff (1874) introduced a + and – notation for designating the configuration of an asymmetric carbon atom. He used mechanical models (built of tetrahedra), and the + and – signs were given by observing the tetrahedra of the mechanical model from the centre of the model. Thus a molecule of the type CabdCabd may be designated + +, – –, and + –. E. Fischer (1891) pointed out that this + and – notation can lead to wrong interpretations when applied to molecules containing more than two asymmetric carbon atoms (the signs given to each asymmetric carbon atom depend on the point of observation in the molecule). Fischer therefore proposed the use of plane projection diagrams of the mechanical models instead of the + and – system. It is important to note here that the Fischer projection formulae are always those of the eclipsed conformations.

Fischer, working on the configurations of the sugar$, obtained the plane formulae (I) and (II) for the enantiomers of saccharic acid, and arbitrarily chose (I) for dextrorotatory saccharic acid, and called it d-saccharic acid. He then, from this, deduced formula (III) for d-glucose. Furthermore, Fischer thought it was more important to indicate stereochemical relationships than merely to indicate the actual direction of rotation. He therefore proposed that the prefixes d and l should refer to stereochemical relationships and not to the direction of rotation of the compound. For this scheme to be self-consistent (among the sugars) it is necessary to choose one sugar as standard and then refer all the others to it. Fischer apparently intended to use the scheme whereby the compounds derived from a given aldehyde sugar should be designated according to the direction of rotation of the parent aldose.

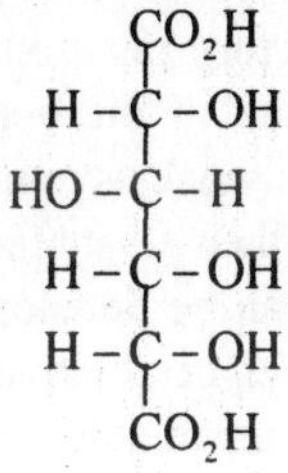
(I)

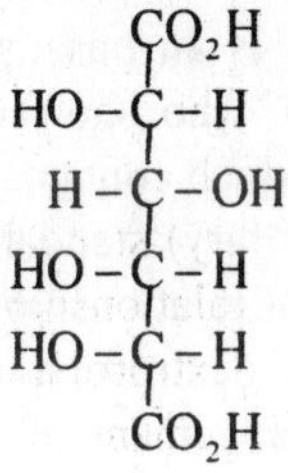
(II)

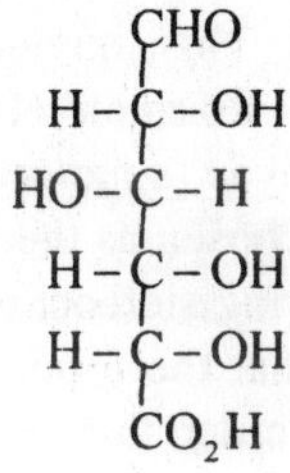
(III)

Natural mannose is dextrorotatory. Hence natural mannose will be d-mannose, and all derivatives of d-mannose, *e.g,* mannonic acid, mannitol, mannose phenylhydrazone, etc., will thus belong to the d-series. Natural glucose is dextrorotatory. Hence natural glucose will be d-glucose, and all its derivatives will belong to the d-series. Furthermore, Fischer (1890) converted natural mannose into natural glucose as follows:

d-mannose → d-mannonic acid → d-mannolactone → d-glucose

Since natural glucose is d-glucose (according to Fischer's scheme), the prefix d for natural glucose happens to agree with its dextrorotation (with d-mannose as standard). Natural fructose can also be prepared from natural mannose (or natural glucose), and so will be d-fructose. Natural fructose, however, is laevorotatory, and so is written as d(–)-fructose, the symbol d indicating its stereochemical relationship to the parent aldose glucose, and the symbol - placed in parentheses before the name indicating the actual direction of rotation.

More recently the symbols d and l have been replaced by D and L for configurational relationships, *e.g,* L(+)-lactic acid. Also, when dealing with compounds that cannot be referred to an arbitrarily chosen standard, (+)- and (–)- are used to indicate the sign of the rotation. The prefixes dextro and laevo (with hyphens) are also used.

Fischer's proposal to use each aldose as the arbitrary standard for its derivatives leads to some difficulties, *e.g,* natural arabinose is dextrorotatory, and so is to be designated D-arabinose. Now natural arabinose (D-arabinose) can be converted into mannonic acid which, if D-arabinose is taken as the parent aldose, will therefore be D-mannonic acid. This same acid, however, can also be obtained from L-mannose, and so should be designated as L-mannonic acid.

Thus, in cases such as this the use of the symbol D or L will depend on the historical order in which the stereochemical relationships were established. This, obviously, is an unsatisfactory position, which was realised by Rosanoff (1906), who showed that if the enantiomers of glyceraldehyde (a molecule which contains only one asymmetric carbon atom) are chosen as the (arbitrary) standard, then a satisfactory system for correlating stereochemical relationships can be developed. He also proposed that the formula of dextrorotatory glyceraldehyde should be written in order that the arrangement of its asymmetric carbon atom should agree with the arrangement of C_5 in Fischer's projection formula for natural glucose.

It is of great interest to note in this connection that in 1906 the active forms of glyceraldehyde had not been isolated, but in 1914 Wohl and Momber separated DL-glyceraldehyde into its enantiomers, and in 1917 they showed that dextrorotatory glyceraldehyde was stereochemically related to natural glucose, *i.e.,* with D(+)-glyceraldehyde as arbitrary standard, natural glucose is D(+)-glucose

The accepted convention for drawing D(+)-glyceraldehyde-the agreed (arbitrary) standardis. The tetrahedron is drawn so that three corners are imagined to be above the plane of the paper, and the fourth below the plane of the paper. Furthermore, the spatial arrangement of the four groups joined to the central carbon atom must be placed as, *i.e.,* the accepted convention for drawing D(+)-glyceraldehyde places the hydrogen atom at the left and the hydroxyl group at the right, with the aldehyde group at the top corner.

Now imagine the tetrahedron to rotate about the horizontal line joining H and OH until it takes up the position. This is the conventional position for a tetrahedron, groups joined to full horizontal lines being above the plane of the paper, and those joined to broken vertical lines being below the plane of the paper.

The conventional plane-diagram is obtained by drawing the full horizontal and broken vertical lines of as full lines, placing the groups as they appear and taking the asymmetric carbon atom to be at the point where the lines cross. Although, is a plane-diagram, it is most important to remember that horizontal lines represent groups

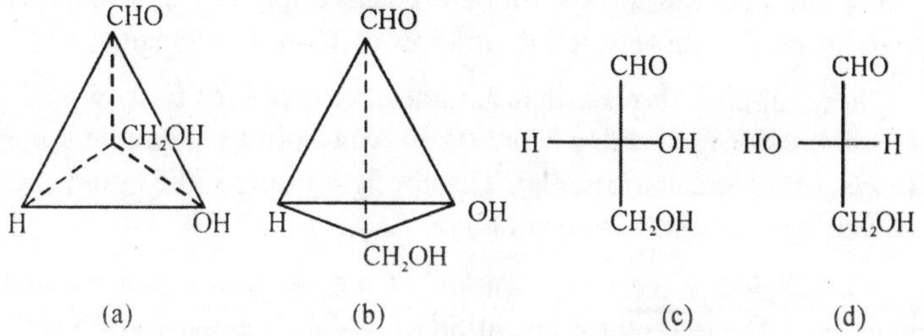

Fig. 4

above the plane, and vertical lines groups below the plane of the paper. Many authors prefer to draw Fig. 4 with a broken vertical line. Fig. 4 represents the plane-diagram formula of L(–)-glyceraldehyde; here the hydrogen atom is to the right and the hydroxyl group to the left. Thus, any compound that can be prepared from, or converted into, D(+)-glyceraldehyde will belong to the D-series. Similarly, any compound that can be prepared from, or converted into, L(–)-glyceraldehyde will belong to the L-series. When representing relative configurational relationships of molecules containing more than one asymmetric carbon atom, the asymmetric carbon atom of glyceraldehyde is always drawn at the bottom, the rest of the molecule being built up from this unit (but see below).

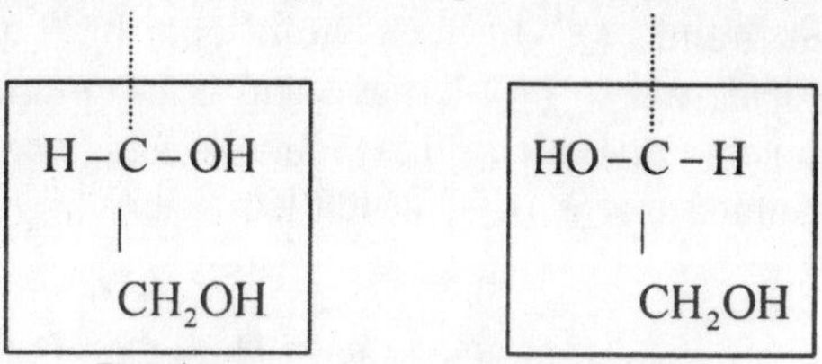

D-series L-series

Fig. 5

Thus, we have a scheme of classification of relative configurations based on D(+)-glyceraldehyde as arbitrary standard. Even on this basis confusion is still possible in relating configurations to the standard (see later).

Until recently there was no way of determining, with certainty, the absolute configuration of molecules. Arbitrary choice makes the configuration of D(+)-glyceraldehyde have the hydrogen to the left and the hydroxyl to the right. Bijvoet *et al.,* (1951), however, have shown by X-ray analysis of sodium rubidium tartrate that it is possible to differentiate between the two optically active forms, *i.e.,* it is possible to determine the absolute configuration of these two enantiomers.

These authors showed that natural dextrorotatory tartaric acid has the configuration assigned to it by Fischer (who correlated its configuration with that of the saccharic acids). The configurations of the tartaric acids, however, are a troublesome problem.

]Fischer wrote the configuration of natural dextrorotatory tartaric acid as (IV). If we use the convention of writing the glyceraldehyde unit at the bottom, then (IV) is L(+)-tartaric

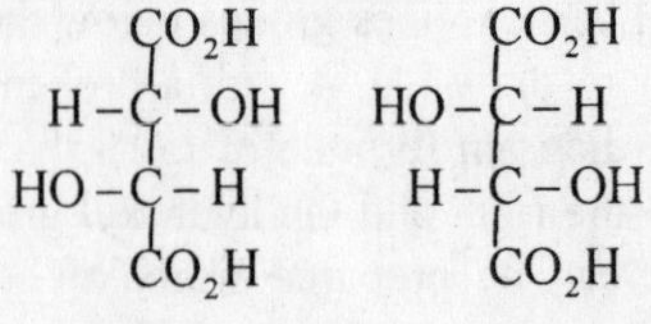

(IV) (V)

acid and (V) is D(–)-tartaric acid. This relationship (to glyceraldehyde) is confirmed by the conversion of D(+)-glyceraldehyde into laevorotatory tartaric acid via the Kiliani reaction. Thus (–)-tartaric acid is D(–)-tartaric acid (V).

On the other hand, (+)-tartaric acid can be converted into D(–)-glyceric acid, and so (+)-tartaric acid is D(+)-tartaric acid (IV). In this reduction of (+)-tartaric acid to (+)-malic acid (by hydriodic acid), it has been assumed that it is C_1 which has been

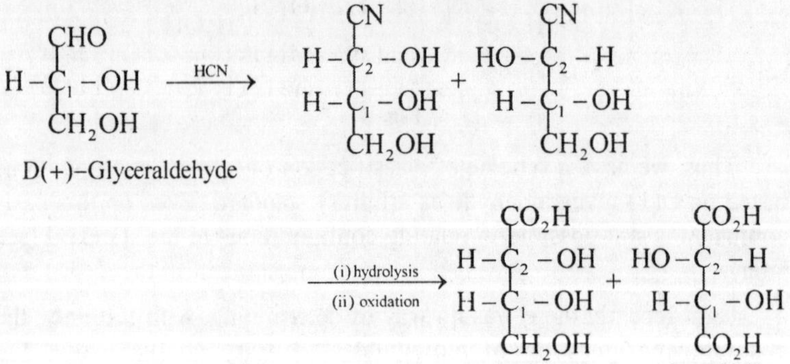

D(+)–Glyceraldehyde

meso–tartaric acid (–)–tartaric acid

reduced, *i.e.,* in this case the configuration of C_2 has been correlated with glyceraldehyde and not that of C_1 as in the previous set of reactions. Had, however, C_2 been reduced, then the final result would have been

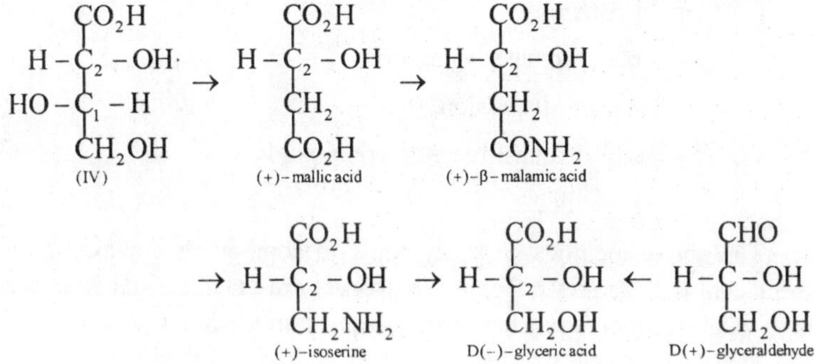

(+)-tartaric acid still through the intermediate, (+)-malic acid (two exchanges of groups give the same malic acid as before). Since (+)-malic acid has been correlated with (+)-glyceraldehyde, (+)-tartaric acid should be designated D(+)-tartaric acid.

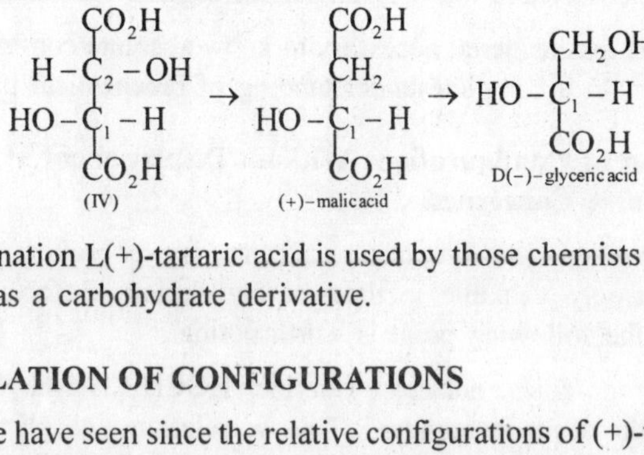

The designation L(+)-tartaric acid is used by those chemists who regard this acid as a carbohydrate derivative.

CORRELATION OF CONFIGURATIONS

As we have seen since the relative configurations of (+)-tartaric acid and (+)-glyceraldehyde have been established, it is now possible to assign absolute configurations to many compounds whose relative configurations to (+)-glyceraldehyde are known, since the configurations assigned to them are actually the absolute configurations. The methods used for correlating configurations are:

(i) Chemical reactions without displacement at the chiral centre concerned.

(ii) Chemical reactions with displacement at the chiral centre concerned.

(iii) X-ray analysis.

(iv) Asymmetric inductive correlation.

(v) Optical rotations:

(a) Monochromatic rotations.

(b) Rotatory dispersion.

(vi) The study of quasi-racemic compounds.

(vii) Enzyme studies.

The above methods normally apply to compounds containing one chiral centre. When several chiral centres are present, the usual procedure is to establish the stereochemistry of the centres relative to each other and then to correlate one of the centres with glyceraldehyde.

It can also be seen that it is better still if it is possible to correlate more than one centre with glyceraldehyde.

Knowledge of absolute configurations is very valuable in the study of optical activity and the mechanisms of organic reactions.

Also, it is considered necessary to know absolute configurations ir order to obtain a complete understanding of biochemical processes.

Correlation of Configurations Without Displacement at the Chiral Centre Concerned

Since no bond joined to the chiral centre is ever broken, this method is an extremely valuable method of correlation. Before discussing examples, the following point is worth noting.

For amino-acids, natural (–)-serine, $HOCH_2CH(NH_2)CO_2H$, was chosen as the arbitrary standard. Thus correlation with glyceraldehyde was indicated by D_g or L_g, and with serine by D_s or L_s. These two standards have now been correlated, and it has been shown that $L_g - L_s$, *i.e.,* natural (–)-serine belongs to the L-series.

The following examples illustrate this method of correlation.

(i) 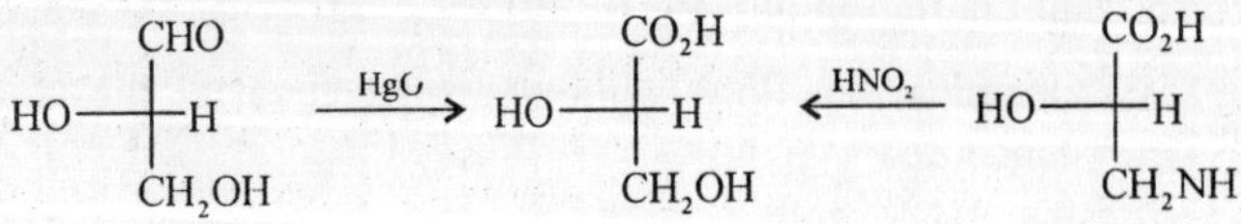

L(–)-glyceraldehyde L(+)-glyceric acid L(–)-isoserie

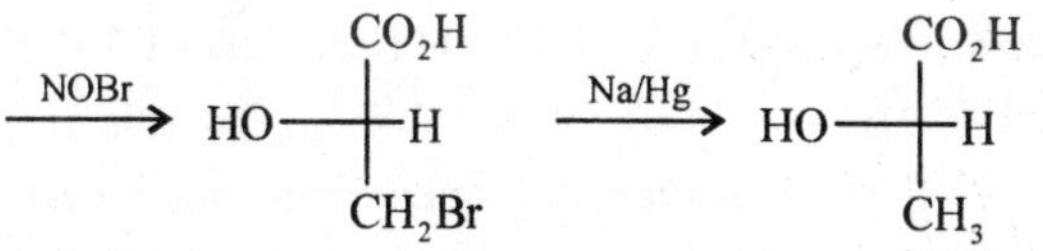

L- L(+)-lactic acid

It can be seen from this example that change in the sign of rotation does not necessarily indicate a change in configuration.

(ii) 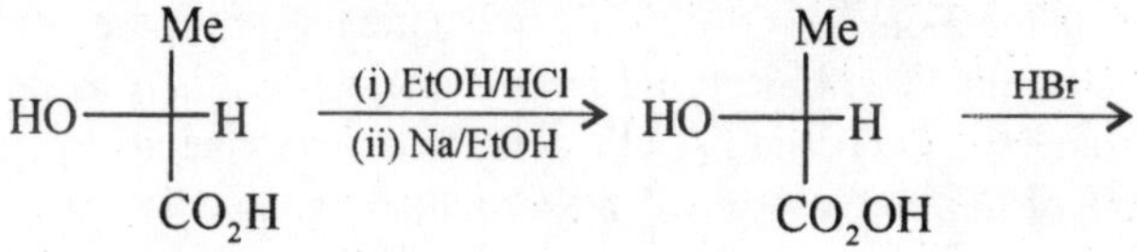

D(+)-lactic acid D-

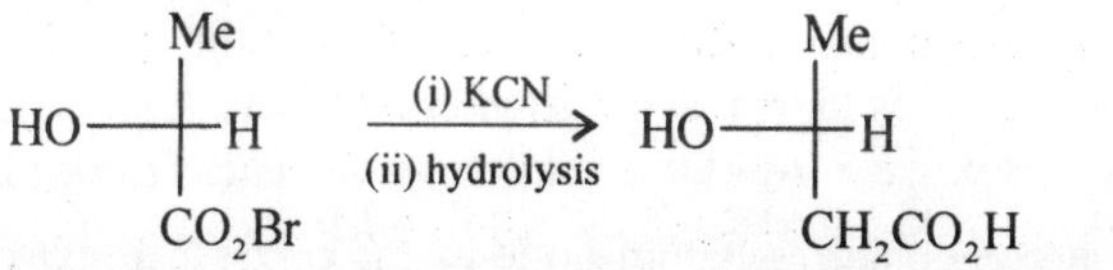

D- D(−)-β-hydroxybutyric acid

(iii) 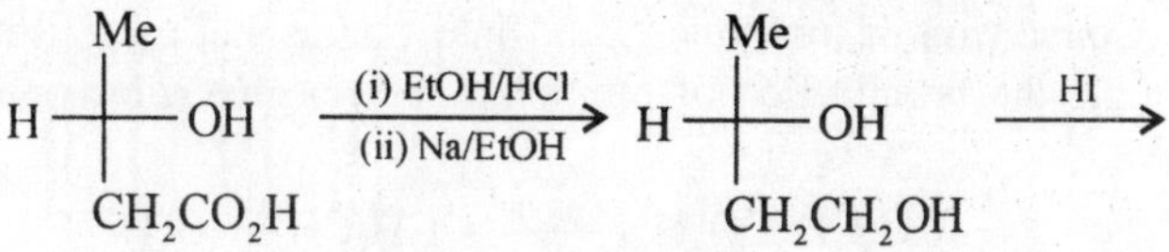

L(+)-β-hydroxy-butyric acid L-

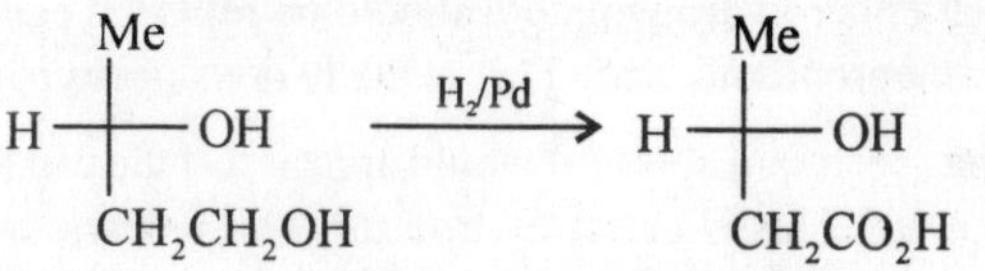

L- L(+)-butan-2-ol

(iv) Another example is that in the terpene series.

(v) All the previous examples are acyclic or alicyclic compounds.

When, however, the compound contains a phenyl group attached to the asymmetric carbon atom, correlation with glyceraldehyde is carried out either by breaking down the phenyl group, leaving only C_1 (usually as CO_2H) attached to the asymmetric.

THE NUMBER OF ISOMERS IN OPTICALLY ACTIVE COMPOUNDS

The number of optical isomers that can theoretically be derived from a molecule containing one or more chiral centres is of fundamental importance in stereochemistry.

Compounds Containing One Chiral Centre

With the molecule Cabde only two optical isomers are possible, and these are related as object and mirror image, *i.e.,* there is one pair of enantiomers, *e.g,* D- and L-lactic acid. If we examine an equimolecular mixture of dextrorotatory and laevorotatory lactic acids, we shall find that the mixture is optically inactive. This is to be expected, since enantiomers have equal but opposite rotatory power. Such a mixture (of equimolecular amounts) is said to be optically inactive by external compensation, and is known as a racemic modification; it is designated as r-, (±)- or DL-, *e.g,* r-tartaric acid, (±)-limonene, DL-lactic acid.

Thus a compound containing one chiral centre can exist in three forms: (+), (–) and (±).

Conversion of molecule Ca_2bd into Cabde. Let us consider as an example the bromination of propionic acid to give α-bromopropionic acid.

$$CH_3CH_2CO_2H \xrightarrow{Br_2/P} CH_3CHBrCO_2H$$

(II) and (III) (Fig. 2.17) are enantiomers, and since molecu!e (I) is symmetrical about its vertical axis, it can be anticipated from the theory of probability that either hydrogen atom should be replaced equally well to give (±)-α-bromopropionic acid. This actually does occur in practice.

From what has been said above, it would appear that the two hydrogen atoms in (I) are alike. This is certainly true for their behaviour towards bromine, but the point to note is that a pair of enantiomers was produced, *i.e.,* replacement of one or the other hydrogen does not produce the same molecule. These two hydrogen atoms are therefore said to be enantiotopic. This term may be defined as follows: Two atoms or groups in a molecule are enantiotopic if replacement of each in turn by some other group leads to a pair of enantiomers. The case of propicnic acid is an example of enantiotopic groups by internal comparison, *i.e.,* the two groups are in the samc molecule. There are also enantiotopic groups by external comparison. Enantiotopic groups of this type are corresponding atoms or groups in a pair of enantiomers, *e.g*, the two methyl groups or bromine atoms in (II) and (III) are enantiotopic. If separate replacement produces the same molecule, the atoms or groups are said to be homotopic.

Prochirality. If a centre in a molecule bears enantiotopic groups, that centre is said to be prochiral. Alternatively, a molecule that contains enantiotopic groups is prochiral, and vice versa. Thus, the carbon atom in (I) is prochiral since replacement of one hydrogen atom by bromine produces a chiral centre. Also, the prefix pro is used to designate a hydrogen (or any ligand) attached to a prochiral centre and the two enantiotopic hydrogens (or ligands) are distinguished by use of the symbols R and S. The symbol to be used is determined by the specification of the chiral molecule produced by replacing a hydrogen atom by deuterium, *e.g*,

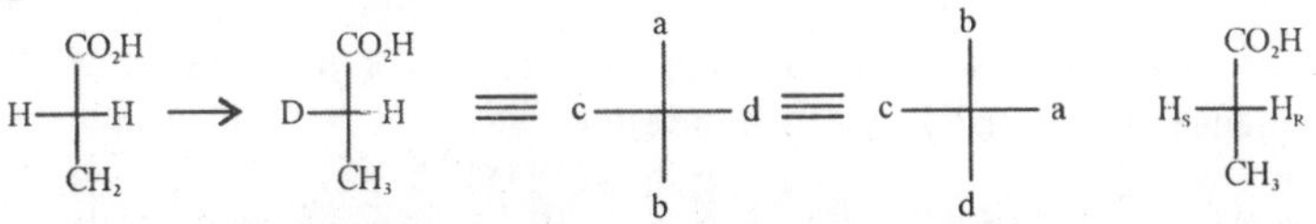

(I) (Ia) (S) (Ib)

The hydrogen replaced in (I) to give (Ia) is therefore pro-S-hydrogen, and consequently the other prochiral hydrogen is pro-R-hydrogen. This may be indicated by writing the formula of (I) as (Ib).

Nuclei which experience equal magnetic shielding have identical chemical shifts; such nuclei are said to be isochronous. Thus, chemically equivalent protons are isochronous, but so are enantiotopic (prochiral) protons, since these also experience equal magnetic shielding, *i.e.,* the signals have the same chemical shifts. However, if dissolved in chiral solvents, then the chemical shifts of enantiotopic (prochiral) protons in a compound may be different.

As we have seen, when (I) reacts with bromine, the pair of enantiomers (II) and (III) are formed in equal amounts. This is due to the fact that when enantiotopic (prochiral) groups react with an achiral reagent, the transition states involved have equal energy contents. On the other hand, if the reagent is chiral, the transition states are diastereoisomeric.

Since these have different energy contents, the two rates of reaction are different, thereby resulting in the formation of a pair of enantiomers in unequal amounts. This may be illustrated by the oxidation of ethanol with the enzyme alcohol dehydrogenase; only one of the two enantiotopic (prochiral) hydrogen atoms is removed to form acetaldehyde. This may be formulated as shown, and it should be noted that the product in this case is not optically active.

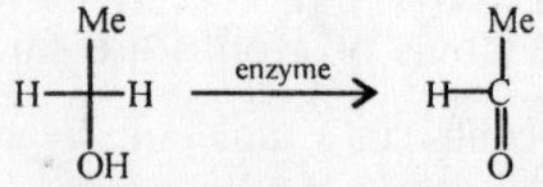

In addition to referring to groups as being enantiotopic or prochiral, faces of double bonds are also said to be enantiotopic or prochiral if stereoisomers are produced by addition reactions, *e.g*, the reaction between acetaldehyde and phenylmagnesium bromide (attack at 'front' and at 'back')

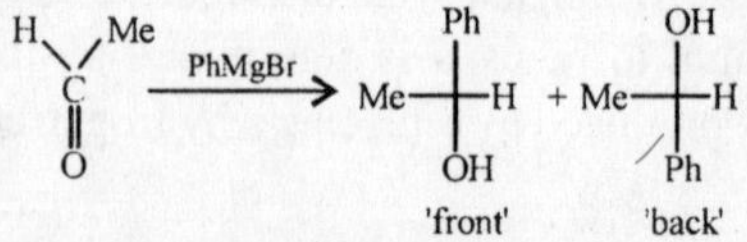

For the purpose of naming the enantiotopic or prochiral faces, the sequence rule is used in two dimensions.

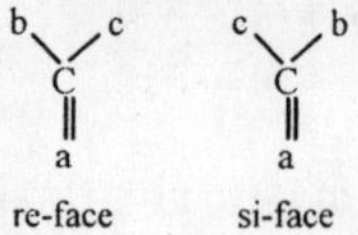

If the order of precedence is a > b > c, then, if the groups are in a clockwise arrangement, that face is called re (rectus), and if in an anticlockwise arrangement, si (sinister).

This nomenclature may be extended to the ethylenic double bond, each end of the double bond being treated separately, *e.g*,

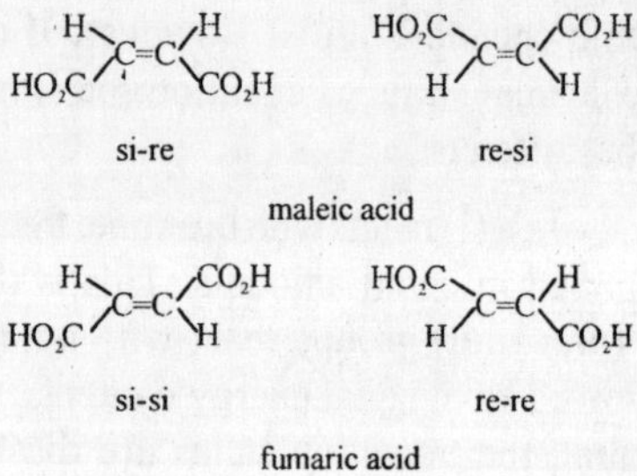

COMPOUNDS CONTAINING TWO DIFFERENT CHIRAL CENTRES

When we examine the molecule Cahd Cabe, *e.g*, α, β-dibromobutyric acid, $CH_3CHBrCHBrCO_2H$, we find that there are four possible spatial arrangements for this type of molecule (Fig. 2.18). (I) and (II) are enantiomers (the configurations of both asymmetric carbons are reversed), and an equimolecular mixture of them forms a racemic modification; similarly for (III) and (IV).

Thus, there are six forms in all for a compound of the type CabdCabe: two pairs of enantiomers and two racemic modifications. (I) and (III) are not identical in configuration and are not mirror images (the configuration of one of the two asymmetric carbon atoms is reversed); they are known as diastereoisomers, *i.e.,* they are optical isomers but not enantiomers. Diastereoisomers differ in physical properties such as melting point, density, solubility, dielectric constant and specific rotation. Chemically they are similar, but their rates of reaction with other optically active compounds are different.

The mass spectra of diastereoisomers may exhibit differences. These differences are usually too small to be significant for acyclic diastereoisomers; this is believed to be due mainly to the fact that these molecules are capable of free rotation about the single bonds joined to the chiral centres. On the other hand, the mass spectra of alicyclic diastereoisomers may differ to such an extent that it may be possible to deduce the stereochemistry of each diastereoisomer from its mass spectrum.

The NMR spectra of diastereoisomers are also different.

The plane-diagrams of molecules (I-IV) will be (V-VIII), respectively, as shown. It should be remembered that groups joined to horizontal lines lie above the plane of the paper, and those joined to vertical lines lie below the plane of the paper.

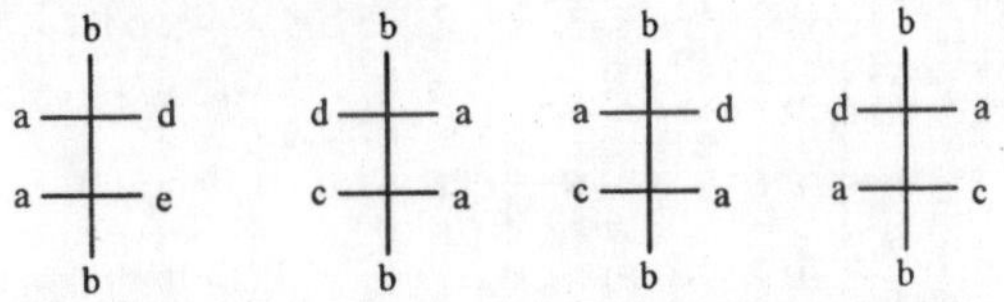

(V) (VI) (VII) (VIII)

Instead of writing down all the possible configurations, the number of optical isomers for a compound of the type CabdCabe may be obtained by indicating the configuration of each asymmetric carbon atom by the symbol + or –, or by D or L; thus;

$$\underbrace{\begin{matrix}+ & -\\ + & -\end{matrix}}_{(\pm)}\quad \underbrace{\begin{matrix}+ & -\\ - & +\end{matrix}}_{(\pm)}\qquad \text{or}\qquad \underbrace{\begin{matrix}D_1 & L_1\\ D_2 & L_2\end{matrix}}_{DL}\quad \underbrace{\begin{matrix}D_1 & L_1\\ L_2 & D_2\end{matrix}}_{DL}$$

Pairs of enantiomers of the type CabdCabe are distinguished by the prefixes erythro and threo. The former is the one in which the identical groups can eclipse each other (a,a and b,b) in one conformation, whereas the latter is the one in which this cannot be done. These names are derived from erythrose and threose, the tetrose sugars. For the relative stabilities of the diastereoisomers.

Conversion of molecule Ca_2bCabe into CabdCabe. Let us consider the bromination of β-methylvaleric acid to give α-bromo-β-methylvaleric acid.

$$CH_3CH_2CH(CH_3)CH_2CO_2H \xrightarrow{Br_2/P} CH_3CH_2CH(CH_3)CHBrCO_2H$$

β-Methylvaleric acid contains one asymmetric carbon atom, but the bromine derivative contains two. Let us first consider the case where the configuration of the asymmetric carbon atom in the starting material is D_1 (IX). Bromination of this will produce molecules (X) and (XI); these are diastereoisomers and are produced in unequal amounts. This is to be anticipated; the two a-hydrogen atoms are not symmetrically placed with respect to the lower half of the molecule, and consequently different rates of substitution can be expected.

In the same way, bromination of the starting material in which the configuration of the asymmetric carbon atom is L_1 (XII) leads to the formation of a mixture of diastereoisomers (XIII and XIV) in unequal amounts. One can expect, however, that the amount of (XIII) produced from (XII) would be the same as that of (X) from (IX) since, in both cases, the positions of the

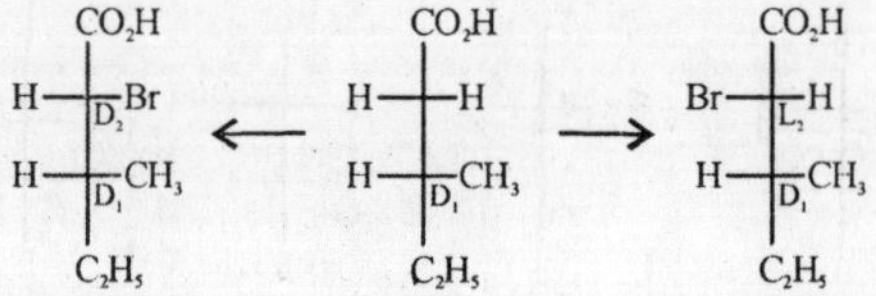

(X) (IX) (XI)

bromine atoms with respect to the methyl group are the same. Similarly, the amount of (XIV) from (XII) will be the same as that of (XI) from (IX). Thus, bromination of (±)-β-methylvaleric acid will result in a mixture of four bromo derivatives which will consist of two racemic modifications in unequal amounts, and the mixture will be optically inactive.

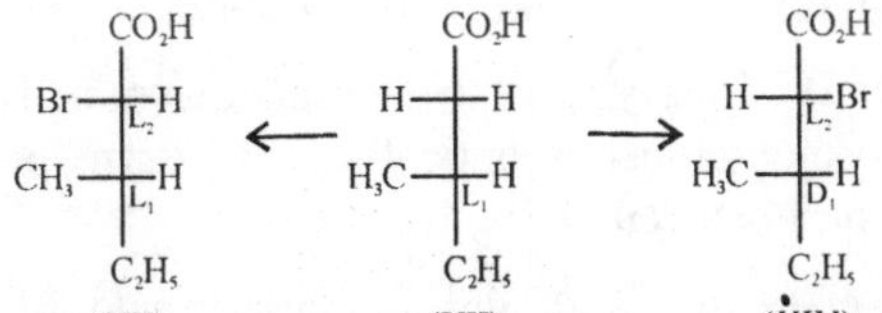

(XII) (XII) (XIV)

As we have already seen enantiotopic groups react with achiral reagents at the same rates. In the molecules (IX) and (XII), however, the two hydrogen atoms react with bromine, an achiral reagent, at different rates to give a pair of diastereoisomers in unequal amounts. These two hydrogen atoms are therefore said to be diastereotopic. This term may be defined as follows:

Two atoms or groups in a molecule are diastereotopic if replacement of each in turn by some other group leads to a pair of diastereoisomers. The molecule we have discussed is an example of diastereotopic groups by internal comparison, *i.e.,* the two groups are in the same molecule. On the other hand, corresponding groups in a pair of diastereoisomers are said to be diastereotopic by external comparison.

The term diastereotopic faces may be used with respect to the faces of a double bond when one of the groups attached to the unsaturated carbon atom contains a chiral centre.

A centre in a molecule which bears diastereotopic groups is also said to be prochiral and the term 'heterotopic' has been used to describe atoms or groups which are not homotopic without differentiation being made whether the atoms or groups are enantiotopic or diastereotopic.

For enantiotopic groups, the transition states for reactions with achiral reagents have the same energy contents, but for reactions with chiral reagents, the transition states are diastereoisomeric and the energy contents are different. For diastereotopic groups, the transition states with both achiral and chiral reagents are diastereoisomeric and consequently the diastereoisomeric products are formed in unequal amounts.

Protons in diastereotopic groups show different chemical shifts (since their environments are different). Such protons are said to be anisochronous, *i.e.,* the signals do not have the same chemical shifts. These different chemical shifts are exhibited whether the solvent is chiral or achiral.

Compounds of the Type Cabd $(Cab)_x$ Cabd

In compounds of this type the two terminal asymmetric carbon atoms are similar, and the number of optically active forms possible depends on where x is odd or even.

(i) **Even series.** (a) CabdCahd, *e.g,* tartaric acid. In a compound of this type the rotatory power of each asymmetric carbon atom is the same. Now let us consider the number of optical isomers possible.

D	L	D	L
D	L	L	D
(I)	(II)	(III)	(IV)

In molecules (I) and (II), the upper and lower halves reinforce each other; hence (I), as a whole, has the dextro- and (II) the laevo-configuration, *i.e.,* (I) and (II) are optically active, and enantiomeric.

On the other hand, in (III) the two halves are in opposition, and so the molecule, as a whole, will not show optical activity. It is also obvious that (III) and (IV) are identical, *i.e.,* there is only one optically inactive form of CabdCabd. Molecule (III) is said to be optically inactive by internal compensation, and is known as the meso-form, and is a diastereoisomer of the pair of enantiomers (I) and (II). The meso-form is also known as the inactive form and has been represented as the i-form; the meso form cannot be resolved. Thus there are four forms possible for the molecule CabdCabd: one pair of enantiomers, one racemic modification and one meso- (i-) form. These forms for tartaric acid are:

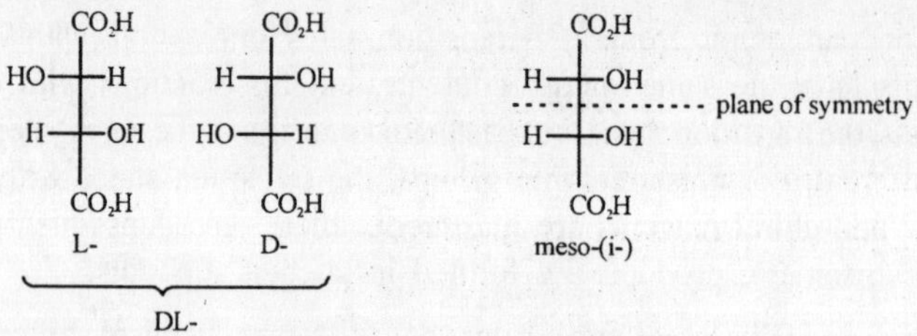

Inspection of these formulae shows that the D- and L- forms do not possess any elements of symmetry; the meso-form, however, possesses a plane of symmetry.

(b) CabdCabCabCabd, *e.g,* saccharic acid,

$HO_2CCHOHCHOHCHOHCHOHCO_2H$

The rotatory powers of the two terminal asymmetric carbon atoms are the same, and so are those of the middle two (the rotatory powers of the latter are almost certainly different from those of the former; equality would be fortuitous). The possible optical isomers are as follows (V-XIV):

D_1	L_1	D_1	L_1	D_1	L_1	D_1	L_1	D_1	D_1
D_2	L_2	L_2	D_2	D_2	L_2	D_2	L_2	D_2	L_2
D_2	L_2	L_2	D_2	D_2	L_2	D_2	L_2	L_2	D_2
D_1	L_1	D_1	L_1	L_1	D_1	D_1	L_1	L_1	D_1
(V)	(VI)	(VII)	(VIII)	(IX)	(X)	(XI)	(XII)	(XIII)	(XIV)
DL		DL		DL		DL		meso-forms	

Molecules (V) and (VI) are optically active (enantiomeric) and are not 'internally compensated'; (VII) and (VIII) are optically active (enantiomeric) and are not 'internally compensated'; (IX) and (X) are optically active (enantiomeric) but are 'internally compensated at the ends'; (XI) and (XII) are optically active (enantiomeric) but are 'internally compensated in the middle'; (XIII) and (XIV) are meso-forms and are optically inactive by (complete) internal compensation. Thus, there are eight optically active forms (four pairs of enantiomers), and two meso-forms.

In general, in the series of the type $Cabd(Cab)_{n-2}Cabd$, if n is the number of asymmetric carbon atoms and n is even, then there will be 2^{n-1} optically active forms, and $2^{(n-2)\backslash 2}$ meso-forms.

(ii) *Odd series :* (a) CabdCabCabd, *e.g,* trihydroxyglutaric acid. If the two terminal asymmetric carbon atoms have the same configuration, then the central carbon atom has two identical groups joined to it and hence cannot be asymmetric. If the two terminal configurations are opposite, then the central carbon atom has apparently four different groups attached to it (the two ends are mirror images and not superimposable).

Thus, the central carbon atom becomes asymmetric, but at the same time the two terminal atoms 'compensate internally' to make the molecule as a whole symmetric (there is now a plane of symmetry), and consequently the compound is not optically active. In this molecule the central carbon atom is said to be pseudo-asymmetric, and is designated 'D' and 'L' (or ⊕ and

⊖ if the + and – convention is used). There will, however, be two meso-forms since the pseudo-asymmetric carbon atom can have two different configurations (see XV-XVIII). Thus there are five forms in all: two optically active forms (enantiomers), one racemic

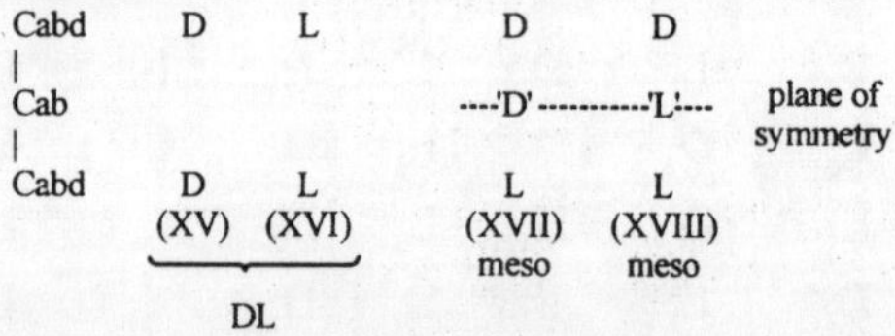

modification, and two meso-forms. The following are the corresponding trihydroxyglutaric acids, all of which are known.

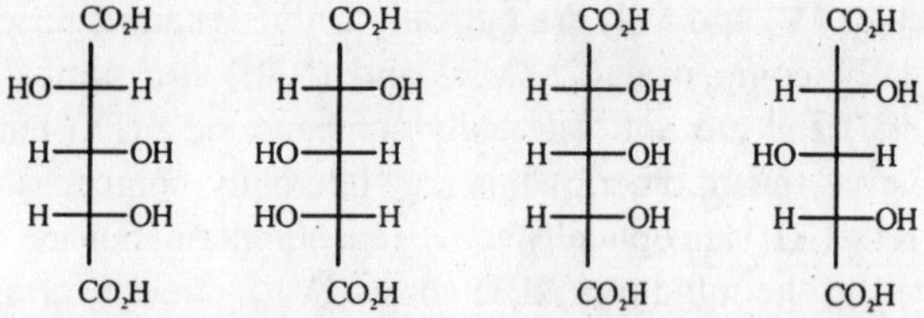

D L meso meso

Inspection of the structure of the trihydroxyglutaric acid shows that the four groups attached to the pseudo-asymmetric carbon atoms are of two types in pairs: one pair consists of two different groups which are achiral (H and OH), and the other pair consists of two groups which are enantiomeric (–CHOHCO$_2$H). These are the characteristics of a pseudo-asymmetric carbon atom, *i.e.*, the molecule is of the type Cabd$_D$d$_L$.

(b) *CabdCabCabCabCabd* : In this molecule the central carbon atom is pseudo-asymmetric when the left-hand side of the molecule has the opposite configuration to that of the right-hand side; the central carbon atom is symmetrical when both sides have the same configuration.

In all other cases the central carbon atom is asymmetric, the molecule now containing five asymmetric carbon atoms. The following table shows that there are sixteen optical isomers possible, of which twelve are optically active (six pairs of enantiomers), and four are meso-forms.

Ends with Opposite Configurations

$$\begin{array}{cccc} D_1 & D_1 & D_1 & D_1 \\ D_2 & D_2 & L_2 & L_2 \\ 'D' & 'L' & 'D' & 'L' \\ L_2 & L_2 & D_2 & D_2 \\ L_1 & L_1 & L_1 & L_1 \\ \text{meso} & \text{meso} & \text{meso} & \text{meso} \end{array}$$

Note the characteristics of the pseudo-asymmetric carbon atom (the central one): two different achiral groups (a and b) and two enantiomeric groups in pairs (D_1D_2, L_1L_2; D_1L_2, L_1D_2).

Ends with the Same Configurations

$$\begin{array}{cc} \begin{array}{cc} D_1 & L_1 \\ D_2 & L_2 \\ \\ D_2 & L_2 \\ \multicolumn{2}{c}{\underbrace{D_1 \quad L_1}_{DL}} \end{array} & \begin{array}{cc} D_1 & L_1 \\ L_2 & D_2 \\ \\ L_2 & D_2 \\ \multicolumn{2}{c}{\underbrace{D_1 \quad L_1}_{DL}} \end{array} \end{array}$$

Molecule with Five Asymmetric Carbon Atoms

$$\begin{array}{cccc} \begin{array}{cc} D_1 & L_1 \\ D_2 & L_2 \\ D & L \\ D_2 & L_2 \\ \multicolumn{2}{c}{\underbrace{L_1 \quad D_1}_{DL}} \end{array} & \begin{array}{cc} D_1 & L_1 \\ D_2 & L_2 \\ L & D \\ D_2 & L_2 \\ \multicolumn{2}{c}{\underbrace{L_1 \quad D_1}_{DL}} \end{array} & \begin{array}{cc} D_1 & L_1 \\ D_2 & L_2 \\ D & L \\ L_2 & D_2 \\ \multicolumn{2}{c}{\underbrace{D_1 \quad L_1}_{DL}} \end{array} & \begin{array}{cc} D_1 & L_1 \\ D_2 & L_2 \\ L & D \\ L_2 & D_2 \\ \multicolumn{2}{c}{\underbrace{D_1 \quad L_1}_{DL}} \end{array} \end{array}$$

In general, in the series of the type $Cabd(Cab)_{n-2}Cabd$, if n is the number of 'asymmetric' carbon atoms and n is odd, then there will be 2^{n-1} optical isomers, of which $2^{(n-1)/2}$ are meso-forms and the remainder optically active forms.

Compounds Containing three Chiral Centres

A molecule of this type is CabdCabCabe, *e.g,* the pentoses, and the number of optical isomers possible is eight (four pairs of enantiomers):

$$\underbrace{\begin{matrix} D_1 & L_1 \\ D_2 & L_2 \\ D_3 & L_3 \end{matrix}}_{DL} \quad \underbrace{\begin{matrix} D_1 & L_1 \\ D_2 & L_2 \\ L_3 & D_3 \end{matrix}}_{DL} \quad \underbrace{\begin{matrix} D_1 & L_1 \\ L_2 & D_2 \\ D_3 & L_3 \end{matrix}}_{DL} \quad \underbrace{\begin{matrix} L_1 & D_1 \\ D_2 & L_2 \\ D_3 & L_3 \end{matrix}}_{DL}$$

All the cases discussed so far are examples of a series of compounds which contain n structurally distinct carbon atoms, *i.e.,* they belong to the series $Cabd(Cab)_{n-2}Cabe$. In general, if there are n asymmetric carbon atoms in the molecule (of this series), then there will be 2^n optically active forms and 2^{n-1} resolvable forms (*i.e.,* 2^{n-1} pairs of enantiomers).

These formulae a also apply to monocyclic compounds containing n different asymmetric carbon atoms; they may or may not apply to fused ring systems since spatial factors may play a part in the possible existence of various configurations.